W9-AUZ-120

VIRTUAL ASCENDANCE

VIRTUAL ASCENDANCE

Video Games and the Remaking of Reality

Devin C. Griffiths

ROWMAN & LITTLEFIELD PUBLISHERS, INC.
Lanham • Boulder • New York • Toronto • Plymouth, UK

Published by Rowman & Littlefield Publishers, Inc.
A wholly owned subsidiary of The Rowman & Littlefield Publishing Group, Inc.
4501 Forbes Boulevard, Suite 200, Lanham, Maryland 20706
www.rowman.com

10 Thornbury Road, Plymouth PL6 7PP, United Kingdom

British Library Cataloguing in Publication Information Available

Library of Congress Cataloging-in-Publication Data
Griffiths, Devin C.
Virtual ascendance : video games and the remaking of reality / Devin C. Griffiths.
pages cm
Includes bibliographical references and index.
ISBN 978-1-4422-1694-5 (cloth : alk. paper)—ISBN 978-1-4422-1695-2 (pbk. : alk. paper)—ISBN 978-1-4422-1696-9 (electronic)
1. Video games. 2. Video games—Social aspects. I. Title.
GV1469.3.G74 2013
794.8—dc23
2013015487

∞™ The paper used in this publication meets the minimum requirements of American National Standard for Information Sciences Permanence of Paper for Printed Library Materials, ANSI/NISO Z39.48-1992.

Printed in the United States of America

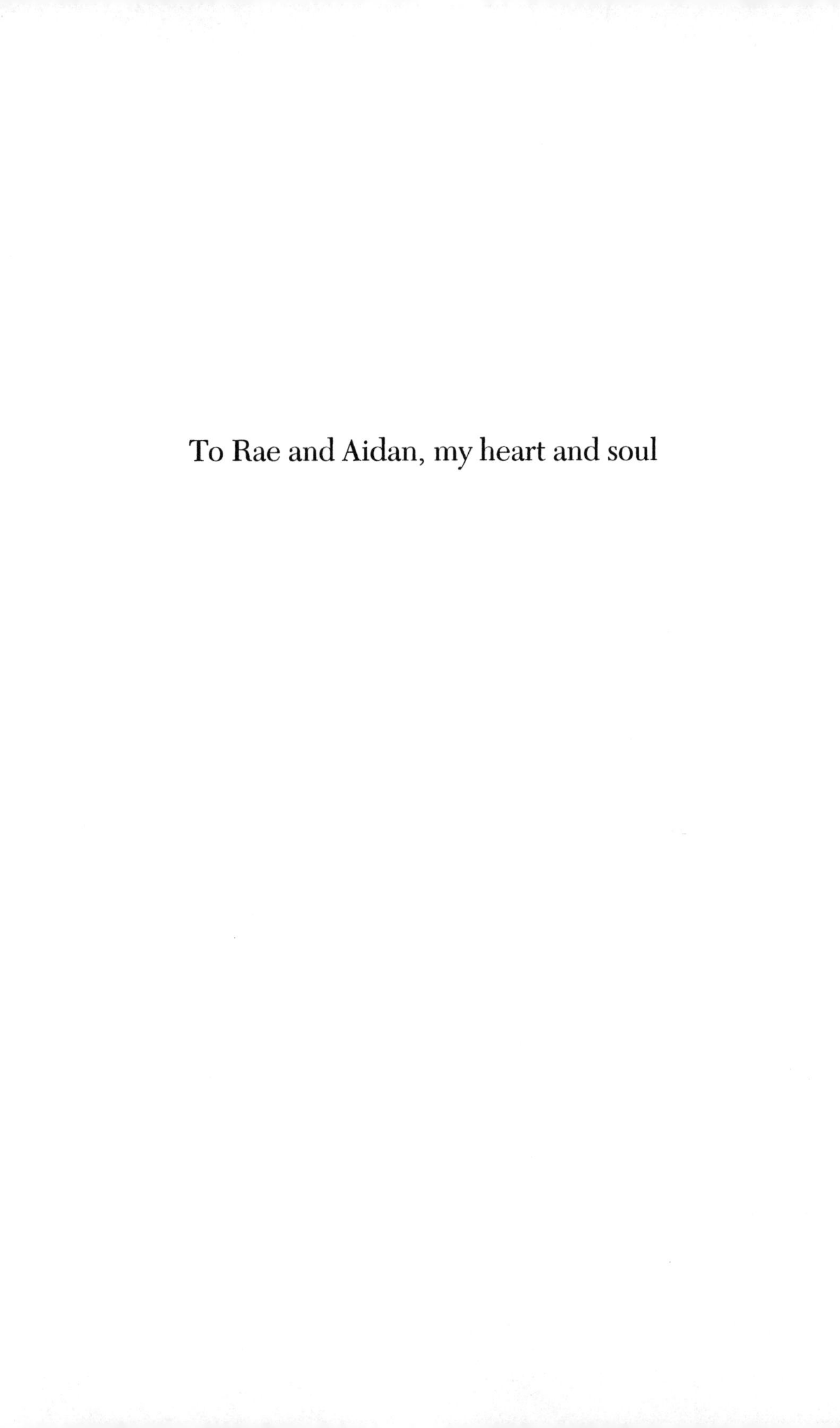

To Rae and Aidan, my heart and soul

CONTENTS

ACKNOWLEDGMENTS

Writing, the act itself, is a lonely process. After you've conducted all the interviews, done all the research, read all the background material, seen all the movies . . . whatever you need to do to actually get *writing*, it all comes down to you and the page.

And it's blank.

Your job as a writer is to fill that page, and all that come after, with everything you've seen, all you've done, everyone you've talked to . . . in short, with the totality of your experience. And the only person you can draw on is yourself. It's your knowledge, your energy, and, hopefully, your skill that comes to bear. You, and you alone, must somehow fashion all the disparate pieces into a coherent, well-crafted whole. Not only is it lonely, it's damn frightening.

But in the process of getting there, no writer is alone—and no book is ever truly a solo effort. In my own experience, there were so many kind, generous, and thoughtful people supporting me that to properly thank them all would take a book in itself. I'll do my best in these few pages, though, with full knowledge that this barely scratches the surface of sufficiency, and with apologies to anyone I may have neglected to mention.

First, thanks to Scott Poulin for setting me on this path (though supplying me with a map would've been nice!), and for being Subject Number One. Next, thanks to all those whose knowledge of the gaming world, enthusiastic willingness to share, and gracious indulgences of often repeated abuses of their time made this entire work possible.

Some of you didn't make it into the book, and for that I apologize. There are more stories to tell than time or space to tell them, at least in one volume. However, everyone I spoke with, whether I was able to include them or not, provided guidance, insight, and inspiration, and I'm grateful to all: Nina Huntemann; Shane Culp; Jerry Heneghan and Steve Cattrell; Nina Fefferman; Attila Ceranoglu; Kent Quirk; Carmen Russoniello; Alex Engel; Andy Brick; Kim Libreri; Rob Lindeman, Tonje Stolpestad, Paulo deBarros, and Jia Wang; Dan Baden; Deb Lieberman; Adam Noah; Bill Crounse; Bob Waddington; Matt Hulett; Mike Arcuri; Peter Hofstede; Chris Wedge; Eleanor Robinson; Mark Barlet; Scott Lambert; Richard Tate; Maria Korolov; Darius Kazemi; Joe Minton; Vincent P. (the Godfath3r); Isaiah "Triforce" Johnson; and Michael Pachter. Special thanks also to the father of video gaming, Ralph Baer, for not only granting me an interview and showing me around his workshop, but for the most enjoyable game of video ping-pong I've played in years.

There were many people behind the scenes who I spoke with or emailed, and who provided incredible logistical support—arranging interviews and press passes, making introductions, providing research material, offering assistance, answering questions, and, where necessary, hand-holding. For all that, I owe a huge debt of gratitude to: Ben Sawyer, Beth Bryant, Emily Burton, and April Matson; Miles Perkins, Taylor Robinson, and Barbara Gamelin; Tina Qunell and Deanna Leung Madden; Lindsey Scott; Donna Austi and Susan Lusty; Jessica Tams; Jen Riley and Ritu Sidhu; David Wang and Christine Nolan; Paul Lambert; Francesca Forrest; Gabriella Mirabelli; Kristin Labriola; Ashley Laurel Wilson; Susan Schwarz; Josh Tolman; John Guerin and Kristina Sampson-Guerin; Dan Lieberman; Steven Solomon; Emily Gabrian; Ann Kuchieski, Nina Mulligan, Mary Senecal, Nancy Bronner, Jennifer Whitehead, and Shelia McCormick at the Clapp Memorial Library; James Peret; Chip Sbrogna; Mike Morasky; Tarnie Williams; Adrian Golub; Shaw Bronner; and Deborah Thomas.

The reality of the publishing world necessitates certain people who many, if not most, would characterize as business associates or, at best, partners. However, the trust, effort, skill, patience, and love that the following people showed in making *Virtual Ascendance* a reality go far beyond that rather formal category. And though I've never met them,

I would be proud to call them friends, and am honored that they chose to work with me. First, my agent, Neil Salkind, who took a chance on me, provided excellent advice and feedback, and worked his tail off (into retirement, as it turns out) to find a great publisher. Thanks also to everyone at Studio B, especially Stacey Czarnowski for carrying on in Neil's footsteps.

A great agent is only one piece of the puzzle, though. A book needs a publisher, and the team at Rowman & Littlefield have been fantastic in shepherding this book from concept to printed reality. Thanks to Suzanne Staszak-Silva, Lindsey Schauer, and Katy Whipple for incredible guidance, care, support, and, yes, patience. You made my first writing experience as painless as possible, and your input has truly improved the book. Also thanks to Devin Watson for an incredible cover design that far exceeds what I imagined.

If it's true that a writer is never really alone, it's doubly true for those with supportive friends and family. Thanks to: Mike Wolpe, Tim Shary, and Jay and Louise Levy for lending ears, keeping me sane, providing advice and encouragement, and for just being great friends; to Scott and Anya Macmillan for supporting me throughout the project and giving me a place to crash; to Georgene Lockwood, for great advice and much help, and for making sure I was well-taken-care of; and to Todd Hahn, for also putting me up, and for a couple of great meals.

In the above-and-beyond category, thanks to Elle Chan, Scott Paterson, and Bee for opening your home to me, introducing me to some really cool people, feeding me, and taking me around San Francisco; to Alana (my favorite cousin), Fred, Avery, and Andrew Black for putting me up and taking care of me in North Carolina; and to Giannina and Dave Silverman for great friendship, company, food, advice, guidance, talking me down, and Batman.

There are two people without whom this, literally, would not have been possible: my parents, Carole and Ed Griffiths. This book is the result of years of your love, encouragement, and support, and it's no exaggeration to say that I couldn't have done this without you. That you're holding this book in your hands is my small way of saying "thank you."

Finally, to Rae and Aidan: There is no language on Earth that has the words to express the depth of my gratitude, my admiration, and my love for you both. You move me and you inspire me, and your patience, support, and love kept me on the path when I often wanted to give up. It's thanks to you that I was able to see this through to the end.

INTRODUCTION

> There are more things between Heaven and Earth than are dreamt of in your mythology.
>
> —*Hamlet*

Three vertical lines and a bouncing square of light. That's all it took to change the world.

If you're too young to remember the Time Before *Pong*, then you probably can't appreciate the momentousness of its arrival. Bear in mind, the game emerged in a very different world. It was a time before home computers, cable television, cell phones, game consoles, the Internet—everything we take for granted today. For many of my formative years, we still watched TV in black and white, and had to get up to change the channel. This was the technological Dark Ages. Had we been less culturally enlightened, we would have denounced *Pong* as witchcraft and burned its inventors at the stake. For those of us who were there—who had never played, let alone seen, a video game—we knew we were witnessing something extraordinary, a groundbreaking achievement in home entertainment. However, none of us knew that we were participating in the birth of a revolution.

From their origins on black and white television sets in homes throughout the country, video games went on to conquer the world. Along the way, they gave rise to or encouraged a horde of technology: graphics accelerators, CD-ROM drives, smartphones and iPads, training simulators, virtual surgical tools, alternate reality goggles, digital movie production, motion capture, DVD and Blu-ray . . . the list is all but endless. If there's a piece of technology you interact with on a daily basis, chances are good that video games had a part in its development.

Today, video games are omnipresent. We play them on dedicated game consoles, computers, laptops, tablets, and cell phones. We play at home, at work, and everywhere in between. We play for hours at a stretch or for a few minutes at a time. We play them alone and, quite often, with others—sometimes in the same room, sometimes halfway around the world. We play for fun, for health, for money, competitively and cooperatively. We play because we can, and we play because we must.

We play.

When I was growing up, video games were all the rage. I spent my youth in the company of games like *Pac-Man*, *Defender*, *Donkey Kong*, *Joust*, and *Lode Runner*, heading out to the arcades or, later, booting them up at home and losing countless hours to escapist fantasy. After playing seriously for fifteen to twenty years, though, I dropped out of the scene in the early '90s for about a decade. When I came back, to my surprise, games had evolved from entertaining distractions to essential tools used in advancing science, health care, training, and education. Professional competitive gaming had been born, and *Dungeons & Dragons*—another favorite pastime of mine—had, through direct descendants like *EverQuest* and *World of Warcraft*, grown into an online obsession for millions.

In January of 2007, I barely knew any of this. But after a chance breakfast conversation with a younger colleague about professional gaming and eBay auctions for items that didn't exist in the real world, I was hooked. This was important. This needed to be told.

Six years later, I'm still at it, though the story I set out to tell has changed. Things that were true a few years ago are no longer, people and organizations have disappeared, while others have arisen in their place. The pace of technological progress has only increased, allowing the realization of ideas today that, when I began this process, were all but impossible—and laying the groundwork for future developments that will, for all intents and purposes, be indistinguishable from magic.

Video games first changed the world with their appearance as an entertainment medium four decades ago. Today, as they address society's greatest challenges and investigate some of our most profound questions, they're poised to do it again. As they advance into the future, they'll begin providing answers and will likely lead us to explore questions that in the present we can't even think to ask.

It's a brave new world that's upon us. We're entering a period of confluence between technology and society the scale of which is unprecedented in our history. Strap yourself in, it's going to be a wild ride.

This is only the beginning.

1

DOWN THE RABBIT HOLE

> I imagine that right now you're feeling a bit like Alice, tumbling down the rabbit hole.
>
> —Morpheus, *The Matrix*

The team moved out from a secure position into the city streets. Four men on a mission: reach their objective, watch their brothers' backs, don't get shot. The threat of violence was palpable, a dead weight pressing on them from all sides. No escape. Inevitability. The attack itself was certain, its timing and direction anything but.

Half the squad moved at a time, two men advancing to a safe point and then covering as the trailing two caught up. Then they'd continue on, sweeping the streets in tandem, clearing corners, checking lines of sight, watching roofs. The enemy could be anywhere.

They worked their way through the city like this, trading places every few blocks, slipping past bullet-pocked buildings and the burnt-out hulks of cars, empty headlight casings staring out like the eyes of the dead. And there, but for the grace of God. . . . You had to forget about mortality, though, ignore the possibility of your own death. Couldn't afford any distractions on patrol. You just had to stay sharp, stay focused, wait for the action.

The attack came suddenly, breaking over the squad like a rogue wave. No warning, no distant sniper fire. Just full-on combat, no holds barred, shockingly fierce. Kills came in rapid fire, death on black beauties. Bang! Man down! This was a one-way ride to hell on the adrena-

line express. Time only for instinct. Shoot first, question later. Cover your man, watch your back, and get your team out alive.

• • •

Sounds hellish, right? A scene out of the latest Hollywood war epic. Or worse, some true-life biopic, a Marine's view of life in a recon unit. Iraq, maybe Afghanistan. Always on edge. Rolling the dice every time you go out. Life and limb. Is today the day your team gets hit? How about tomorrow? Next week? You can prepare all you want, but it's still a surprise when it happens. And right on the heels of "I'm in it now" comes the question, "but am I gonna make it out?"

Unless you're standing where I am: The Meadowlands Expo Center, New Jersey. April. It's Major League Gaming's season opener, and ten thousand hardcore video gamers, split into teams of four, are gearing up for three days of intense competition. Like a great host of predatory birds—sleek, fast, deadly—they descended on the convention center floor to destroy each other on virtual fields of combat. *Halo 2*, *Gears of War*, *Rainbow Six: Vegas*. Each game had its own version of the chapter's opening scene, and over the next seventy-two hours, I'd watch it play out repeatedly, with minor variation, as teams strove to eliminate one another from the event and move a step closer to claiming the winner's purse. The prize? Five, maybe ten thousand dollars and a slot on the next stop of the professional video gaming league's annual showdown of the sport's best.

That's right: *professional* video gaming. Played by pro cyberathletes. With real sponsors—from gaming-specific companies like Turtle Beach, Mad Catz, and Sony PlayStation to household names like Samsung, Dr. Pepper, and Bic—for real money: win a national championship, and your squad of four can walk away with a cool hundred grand, simply for beating another team at a video game. And while that may seem unbelievable, it's only the beginning, our first tentative step into the rabbit hole. Pro gaming is just one facet of an industry that stretches the idea of what's possible and blurs into indistinction the line separating virtual from real.

But I'm getting ahead of myself. We've got some ground to cover before we can explore this or any of the video game industry's other bizarre offshoots and impacts. Before we set out on our journey, let's consider something more familiar. Let's look at icebergs.

As a kid, icebergs fascinated me. Those mammoth hunks of glacier silently patrolling waters so cold they'd turn you into a human Popsicle. Crystalline killers poised to send unwary craft down into the icy depths. Merciless, indestructible humblers of man. Outside of great white sharks, the oceans held nothing more terrifying, and I was convinced that, were I to voyage seaward, my trip would end in ice-borne catastrophe. But what really captivated me was that the piece you could see was only a small fraction of the whole. The rest of the iceberg—the dangerous part—lay invisible beneath the surface, a hanging mountain of frozen menace. Unless you understood icebergs, unless you knew what you were looking at, you'd never realize that what you perceived was an incomplete representation of their reality.

Video gaming is exactly the same.

Most of us view video games as diverting stress-relievers or high-tech babysitters at best. At worst, they're mindless, socially isolating incubators of violence and depravity. And gamers? They're pasty, overweight, awkward young men whose only successful relationships are with PCs and Xboxes. Our dominant cultural view is that nothing in the realm of gaming offers anything positive beyond entertainment—and even that's questionable. This couldn't be further from the truth.

Now, I've spent more hours than I care to admit behind coin-ops, consoles, and computers playing a whole host of games, and I *still* unconsciously bought into much of the accepted cultural wisdom. My mistake. One discussion with an amateur gamer knocked my previous concepts of gaming out cold; my first big gaming events—Major League Gaming's 2007 pro season kick-off and the fifth annual Games for Health Conference two years later—put the nail in the coffin. Based on those experiences alone (never mind the ones I've had since), I can tell you this: our societal stereotypes about video gaming have little to do with reality. Women play games. And men over twenty. Parents play—with and *without* their kids. I've spoken to and seen more varieties of gamers than I'd thought possible. And very few of them were overweight. Yes, the gaming cave troll exists, but it's an endangered species, pushed to the brink by a variety of more adaptable forms. The truth is this: like dragons and unicorns, the typical gamer is a mythical beast. And as I've learned, the impact of video games extends well beyond their traditional domain of entertainment, directly or indirectly touching all aspects of our lives. Business, economics, health, education, medicine, science—even love and war—have all been affected by video gaming. This is the most powerful agent of change to emerge since television.

If you were conscious at any point during the last four decades, then you know something about video games, or you've at least heard of them. They're practically unavoidable: Hollywood's made movies about them since the '80s, video games and virtual worlds have featured heavily in episodes of TV shows like *The X-Files*, *CSI*, and *Law & Order*, and the news media has had a field day vilifying them from day one. No matter the depth of your experience, though, it's unlikely you know the full story. Even professional gamers who spend entire working weeks training, practicing, and honing their skills are unaware how wide the video game industry's net reaches.

The economics alone are staggering. Consider this: despite the Great Recession and faltering US recovery, The Entertainment Software Association's *Essential Facts for 2012* reported the video game industry's 2011 domestic revenue at nearly twenty-five *billion* dollars[1] —just slightly off its 2010 record high of $25.1 billion. That's more than twice Hollywood's own 2011 haul of $10.2 billion for both the United States and Canada,[2] and better than year-end revenues for Major League Baseball, the National Football League, National Basketball Association, and National Hockey League *combined*.[3]

There's far more to the story than just numbers, though. Researchers are exploring the power of video games to heal: Yale University's Dr. Adam Noah is demonstrating *Dance Dance Revolution*'s ability to treat symptoms of Parkinson's disease; led by Dr. Sally Merry, a team of New Zealand psychiatrists is building traditional cognitive behavioral therapy into a video game to help teenagers deal with depression[4]; and a growing number of health care professionals are using video games to help people lose weight, manage diabetes, conquer addiction, treat high blood pressure, and combat diseases like Alzheimer's and childhood cancer. The US Department of Defense is even looking at the ability of video games to help soldiers recover from post-traumatic stress disorder and traumatic brain injury. And by combining the Internet's reach with the power of crowdsourcing, scientists are getting the average citizen involved: In the game *FoldIt*, players around the globe compete online to fold proteins into their optimal shapes, which is devilishly hard to do in a lab, but relatively simple in a computer simulation. Scientists can then work the virtual solutions into real-world labs; the results may lead to drugs for treating some of our most devastating illnesses.

Games have infiltrated the areas of education and training as well. Students enrolled in Florida Virtual School—a fully accredited, Internet-based public high school with more than 120 course offerings—immerse themselves in US history through the 3D adventure *Conspiracy Code*. Developed by subject matter experts and academic scholars, the course covers topics from the American Revolution through US involvement in the Middle East and the war on terror—all experienced within an interactive virtual learning environment. Meanwhile, a University of Wisconsin study of *World of Warcraft* players revealed a penchant for scientific thinking: The game, it seems, encourages players to use systems, models, math, and testing to understand situations and solve problems.[5] By playing *World of Warcraft*, online gamers are unintentionally creating a virtual environment that teaches real-world skills. In the world of Formula 1 racing, professional drivers prepare for upcoming races by taking test runs in highly accurate F1 simulators—like those at Virgin Racing, Red Bull Racing, and Ferrari, which can display any track on the Formula 1 circuit and respond to track conditions exactly like the real cars. Simulators can also help save lives: In 2009, Halifax neurosurgeon Dr. David Clarke made history when he became the first person to remove a brain tumor less than twenty-four hours after removing the same tumor virtually on a 3D simulation of his patient.[6] More surprising is the discovery by researchers at Iowa State University and New York's Beth Israel Medical Center that surgeons who play off-the-shelf video games perform better in the operating room[7]—so much so that developers are now designing low-cost software specifically for surgical training. In the very near future, medical schools with shallower pockets will finally have an effective alternative to the expensive, high-tech training devices beyond their reach.

Of course, all is not sweetness and light. For a small percentage of gamers (most research puts it at 10 to 15 percent), video game addiction is a serious problem, similar to gambling or substance abuse. Pathological gamers suffer withdrawal symptoms and intense cravings, lose interest in other activities, become socially isolated, experience significant weight gain or loss, lie about the amount of time they spend playing—in short, they behave much like typical addicts. Though not yet recognized as a mental health condition by the American Psychological

Association or the American Medical Association, both organizations agree that it bears further study. They stress, however, that it has more to do with individual players than the games themselves; like other addictions, it appears to be a symptom of a more serious underlying issue. Also, for the many benefits they provide, virtual worlds and online role-playing games are not immune to society's dark side: *Second Life*, *MapleStory*, and *World of Warcraft* enthusiasts can become victims of racism, sexism, and harassment, which are particularly insidious in this medium as the perpetrator's true identity is often hidden. I've experienced a little of this myself, having picked up one or two griefers (in-game bullies, in the lexicon of video games) during my stint as a Don in Zynga's Facebook-based *Mafia Wars*—more annoying than harmful, in my case. At the extreme, though, people have in-game affairs (often dooming their real-world relationships), vandalize digital property, and occasionally murder each other's online personas (avatars)—again leading to real-world consequences, as in the 2008 case of a woman arrested for doing in her virtual husband. The potential sentence? Five years in jail and a five thousand dollar fine.

Video games have also attracted the military's attention for more than just rehab: they're powerful training and recruitment tools, and allow extension of the military brand to younger—and more impressionable—kids. *America's Army*—by all accounts a great game, and one of the most consistently popular first-person shooters on the market—was created for exactly that, immersing tweens and teens in a captivating simulation of life in the US Army. And on the frontlines, soldiers use video game–style interfaces to control unmanned bombing and reconnaissance vehicles; while this keeps more of our servicemen and -women out of harm's way, it distances those at the controls from the destruction they create and sanitizes the business of killing, potentially making it easier to carry out.

Food for thought, certainly, but it's all just scratching at the surface. Like any subject, the good stuff lies within, and to reach it, you have to dig. It helps to have a starting point, though, and for us, there's no better place to begin than by exploring the concept of gaming.

I AM, THEREFORE I GAME

Gaming is an essential human activity. From the bygone backgammon contests of ancient Rome and India to friendly chess and checkers matches occurring regularly in city parks and public spaces today, every society since the dawn of recorded time has created and played games. Wei Hai, the oldest known war game, became popular in China around 3000 BC, Iranian and Egyptian excavation sites have yielded up dice and Senet boards older still, and though there's no direct evidence, some aspects of Mancala—which requires no specialized equipment—bear surprising resemblance to human agricultural practices, suggesting that it may be the oldest "board" game ever played. It seems likely that games date back even farther, perhaps to the beginnings of humanity

In his 1938 treatise *Homo Ludens* (which translates to *Man the Player*, or *Playing Man*), Dutch historian and writer Johan Huizinga argues that play is the primary force driving human cultural evolution and actually bestows upon it a higher level of importance than sustenance, sex, or survival:

> As a regularly recurring relaxation, [play] becomes an integral part of life in general. It adorns life, amplifies it and is to that extent a necessity both to the individual—as a life function—and for society by reason of the meaning it contains, its significance, its expressive value, its spiritual and social associations, in short as a culture function. . . . It thus has its place in a sphere superior to the strictly biological processes of nutrition, reproduction and self-preservation.[8]

Of course, we've long known about the biological and social value of play, but Huizinga attaches to it a special significance: Human civilization, he argues, arises from play and is impossible without it.[9] Combine this idea with the ubiquity of games, and gaming seems less an activity for casual entertainment and more a necessity for our survival. Like worrying about the future and obsessing over the past, engaging in gameplay is fundamental to our experience as human beings. Video games are just a natural extension of this phenomenon, the next step in the evolution of gaming.

So, to be human is to game. What does this mean? Just this: you're a gamer. Period. Skeptical? Have you ever played *FreeCell* on your PC?

How about *Scrabble*? *Tetris*? *Bejeweled*? Ever played a Facebook game or done a Sudoku puzzle online? Maybe a game app on your iPod, tablet, or smartphone is more your speed. Regardless, choice of game is beside the point. The issue is the difference between the *perception* of the gamer and the *reality* of gaming. You don't need to be hardcore and spend hours at a stretch attached to a machine for the label to apply: Gaming encompasses far more than *Halo*-style shooters and online role-playing games like *World of Warcraft* or *EVE Online*. Yes, dedicated gamers have an extensive menu of first-person shooters, side-scrollers, sports games, and role-playing games to choose from, and they spend countless hours hacking, blasting, and fighting their way through them. But if you've ever killed time with less involved games, you fall into the category of *casual* gamer—the largest group, by far, and the fastest-growing segment of the gamer population—and for you, there are virtually limitless opportunities to forego sleep or avoid work, online and off, in easy-to-digest five-minute doses.

If you're under forty, chances are you already know this and are comfortable with it. However, if you're like me—over forty and living in the shadow of the 1980s gamer geek label—you may hesitate to admit your gamer status to all but your closest family and friends. Well, you can relax. Gaming has moved from the fringe to the center. And it's *cool*. From school kids regaling each other with the weekends' greatest gaming moments to the White House Press Secretary dropping video game references during a press conference, video games are now an integral part of our popular culture. There's no escaping it: We're a nation of gamers.

That being said, you may still insist that you're not one of *them*, that you've never played a video game in your life. Okay, granted. It's possible. However, you've played *some* sort of game—poker, bridge, *Monopoly*, take your pick—and so, by definition, that still makes you a gamer: The urge to play games in any format comes from the same place. And even *if* you haven't played video games, if you've ever browsed the Internet, viewed a digital photo, created a PowerPoint presentation, or used a computer to draw anything, video games have changed your life: Your PC's fleet-footed graphics chip evolved as the result of a cutting-edge video game's need for speed. And that tablet or touch-screen cell phone you're so fond of? It's a direct descendant of a handheld video game.

You may have felt ripples from the gaming sector's impact on the overall US economy as well: According to the Entertainment Software Association's *Video Games in the 21st Century: The 2010 Report*, the entertainment software industry added $4.9 billion to the US gross domestic product in 2009. Between 2005 and 2009, it grew 10.6 percent—against a mere 1.4 percent real growth for the nation's economy as a whole. Even during lean times, by the end of 2009, the US game industry cleared twenty billion dollars in revenue.[10] December 2009 alone broke all records, ringing in at $5.5 billion and closing the biggest month in the industry's history. And in the two years since the Entertainment Software Association created that report, the industry posted revenues of $25.1 and $24.75 billion (2010 and 2011, respectively).[11] Gaming is far from its deathbed, and by most accounts seems poised to continue record-setting growth into the foreseeable future.

Now before you accuse me of viewing games through rose-colored glasses, my goal isn't to glorify video games or the industry behind them. Whether they're good or bad is irrelevant: Video games simply *are*, and they're not going away anytime soon. Once we accept this, we can move beyond the hype and hysteria and explore how video gaming reflects, meets, drives, and responds to our society's needs, changes, and desires. And we can take a look at what video games really *are*, what they can do now, and what they offer in the future. Yes, they can provide hours of entertainment (sometimes more than you intend: on several occasions I've played past sunrise), but there's more to it than that. At their best, they can teach us new skills, connect us to each other, keep us healthy, and possibly save our lives. Ultimately, by dealing with their reality, we can determine how video games affect our continuing cultural evolution.

This is more than a story about video games, though. It's the story of an awakening, of a realization that while I wasn't looking, a childhood pastime popular with lonely hearts and geeks exploded into a thriving, multibillion dollar enterprise—one rooted in entertainment but whose tendrils reach into virtually all aspects of society. And it's an adventure into a bizarre and shadowy realm where things are rarely what they seem. So put on some comfortable shoes, pack a snack, grab your flashlight, and take a trip down the rabbit hole with me into a land where the line separating real and unreal vanishes. We'll explore this virtual Dark Continent together, and in doing so, come face-to-face with the question at the heart of our relationship to it: In the end, what is reality, and how do we define it?

Ready to go? Great. First stop: leg warmers, acid wash, and big hair. We're going back to the '80s.

2

FROM THE COIN-OP TO THE CONSOLE

How Did We Get Here?

The parking lot is packed, but you're lucky and find a spot not too far from the entrance. As you get out of the car, you can hear the driving beat of '80s glam-rock through the closed door of the building a few dozen feet away. Your pulse quickens as the first hints of adrenaline hit your bloodstream. Crossing the parking lot, you reach into the inside pocket of your faded jean jacket, closing your hand around the rolls of quarters resting there like shotgun shells. Good. For the next few hours, that's all you'll need. You steel yourself against the imminent audio blast wave and pull open the arcade door. The familiar shock of sound hits you, and you crack a smile: a concert of electronic beeps, boops, blasts, blips. The roar of engines and the thunder of explosions. Voices raised in triumph and defeat. And over it all, the steady pulse of the '80s best blaring from the CD jukebox. You make a quick circuit around the arcade floor, watching your peers pit themselves against a spectacular array of video game villains, legs apart, eyes locked, fingers flying across the controls like Mozart at the piano. The latest incarnation of the timeless struggle between man and machine. Like their pinball-wizard forbears, the best players draw crowds: fanboys, groupies, wanna-bes. You give them a courtesy glance but keep moving. Empty machines call out to you like sirens as you pass by, but you resist their seductive pull. You're on a mission, a quest. Your favorite machine. Your game.

There.

You greet it like an old friend, line up a row of quarters on the cabinet—this is your territory now—and slip behind the controls like you're slipping into a well-worn pair of jeans. As you drop the first coin into the slot, your adrenaline spikes to full fight-or-flight. Ready player one. Game on.

• • •

Growing up in the '80s, there were certain things you could count on: aerobic leg warmers, the arms race, MTV. And, of course, video games. They were everywhere. Sure, you had your video arcades—and I spent more time in them than I care to recall—but it went far beyond that. Hotels and airports had dedicated gamerooms (Orlando International and Miami's Fontainebleau Hotel were perennial favorites). Corporate business conferences featured hospitality rooms stocked with video games. Chains like Nathan's and Chuck E. Cheese's sprang up around arcades, offering cheap, fast food to gamers whose only culinary requirement was speed—stepping away from the games just long enough to wolf down a hot dog or slice of pizza. Movie theaters, restaurants, bars, bowling alleys, department stores—all reserved space for the invasion of the machines.

My regular haunt lay on the edge of route 119 in White Plains, New York: White Plains Bowl, a short ride on bus 13B from my home in Tarrytown. I'd head out there whenever I could, armed with stacks of quarters and ready to feed them rapid fire into the waiting machines, like ammo into some high-tech weapon of the future. Locked and loaded, I'd lose myself in the graphic frenzy, shutting out everything around me, letting imagination draw me through the glass screen and into the heart of the action, searching for that glorious moment of union between boy and machine, fingers moving of their own accord, driven by the very electrons emanating from beneath. Sometimes I'd play so long I'd sweat. Quarters spent, I'd step away from the machine, vibrating with resonant energy like a caffeine junky after a serious bender. I wouldn't say I was addicted. But I'd always come back for more.

The arrival of the video game was a pivotal moment in my personal history, coming as it did during my awkward tween/teen years when being outside the social circle wasn't a choice, and when my imagination, at least, ran wild. A few years earlier, *Star Wars* had hit the big screen, and overnight, George Lucas changed the world for an entire industry and an entire generation of kids simultaneously. For a shy young boy, there was nothing better. I wanted to be Han Solo, to fly the Millennium Falcon through a hail of TIE fighters, to dogfight over the Death Star. And I lusted after Princess Leia. Video games may not have gotten me the princess, but for everything else, they were perfect.

Now, I'm talking not about the *technological* arrival of video games—they'd been around years before the advent of the arcades—but of their emergence as a *cultural* phenomenon—marked predominately by the rise of the video arcade and the subsequent eruption of home game consoles, when video games exploded into the popular consciousness with a force matched only by the overcharged hormonal rush of the young males who became their biggest fans.

From their inception, video games progressed along two main parallel paths. The first began with *Pong*, wound through 1980s arcades in various incarnations like *Pac-Man*, *Defender*, and *Tempest*, and evolved into games like *Counter-Strike*, *Quake*, *Halo*, and *Gears of War*—first-person shooters, player versus player, team versus team, or player versus machine. The second has its roots in that favorite pastime of '80s game geeks everywhere, *Dungeons & Dragons*—which directly inspired William Crowther's *Colossal Cave* and eventually spawned role-playing games (RPGs) like *Ultima*, *EverQuest*, *EVE Online*, and *World of Warcraft*. An extension of the second path grew out of an activity that predates video gaming by centuries, if not millennia: pretend play. The human ability to playfully imagine and create is the source of every simulation game on the planet and the reason for their success. *SimCity*, *Spore*, *Outpost*, *Zoo Tycoon*, they all take the idea of pretend and extend it into the world of gaming—ultimately resulting in virtual worlds like *Second Life*, *MapleStory*, *Whyville*, and *The Sims Online*. Virtual worlds are essentially role-playing games, but they move beyond objective-based gaming to a hybrid of gaming and real-world interaction. This lends them more of a broad cultural and economic impact, and as such, they can't *really* be called games (not if we're being honest, anyway).

In recent years, another flavor of video games has risen to prominence: casual games. Quick to learn, easy to play, casual games are aimed at the masses. Casual games have several ancestors, from traditional games like chess and poker, to the Sunday crossword, to RPGs, shooters, puzzles, strategy games—the list goes on. They can be played for hours at a stretch or during a five-minute break between meetings, and they appeal to people of different ages, genders, backgrounds, and interests—most of whom are not hardcore gamers. Driven largely by the arrival of social games (those found on Facebook, for example) and the explosion of mobile devices like smartphones and tablets, casual gaming is the largest, most profitable, and fastest-growing genre in the entire industry.

We'll delve into these different areas in later chapters. For now, let's journey back in time to witness the origin of the video game and explore how we got from then to now.

LET THE GAMES BEGIN

August 5, 2008. Waltham, Massachusetts. The Skellig, Irish pub and site of Boston Post Mortem—the monthly gathering of Boston-area video game developers. The house was packed. One hundred fifty, maybe two hundred people, to see the unveiling of *Rock Band 2*. Weezer. Skynyrd. Talking Heads. A little Jimmy Buffet (who, I'm told, rerecorded a selection of songs specifically for *Rock Band 2*—including a shout-out to the game itself. He's a big fan, I hear). People jump onstage, grab instrument-shaped controllers, and form impromptu bands to play at rock stars, basking in the audience's cheers and adulation.

And then a wizened old man and his equally ancient game console take the stage. The machine is the Brown Box. Prototype of the Magnavox Odyssey, sister to a Smithsonian resident. The man? Ralph Baer. Engineer, inventor, and father of the video game. He's the reason they've come: Like religious pilgrims, the faithful have gathered to behold the man who, more than four decades ago, unwittingly gave birth to their entire industry.

The story begins in 1951. The initial video game spark flared amid a surge in post-war television production. As with the origin of most great inventions, it was unintentional: Quick to recognize the potential of this emerging market, New York electronics company Loral asked one of their engineers to design the best TV set ever seen. If the engineer had been anyone other than Ralph Baer, he might simply have stayed within the bounds of the task, designed the set and left it at that.

But he didn't.

Baer was intensely curious, and the standard procedure for testing new TV sets—using a piece of test equipment that allowed you to manipulate patterns and lines on the screen—gave him an idea: If you could build this functionality directly into the set and allow viewers to control, say, dots on a screen, you could transform TV from a passive to an active experience. People might even be able to play games on it. But would they want to?

It would be awhile before he found out. Loral, already behind production schedule, had no interest in pursuing Baer's radical and untested vision—and their chief engineer was completely underwhelmed by it anyway. And that was that.

Until August 31, 1966. Baer had left Loral eleven years earlier and was now running the Equipment Design Division of Sanders Associates, a military electronics contractor headquartered in New Hampshire. In a rare moment of downtime during a business trip to New York, while waiting for a colleague, his fifteen-year-old TV gaming idea resurfaced—and this time, Baer didn't let it go. He grabbed a notepad and wrote like a man possessed. The next morning, September 1st, Baer expanded his notes into a four-page document that laid it all out. Though he couldn't possibly have predicted it, he'd just written the Book of Genesis for the entire gaming industry. From there, Baer shanghaied fellow Sanders technician Bill Harrison to help build a prototype, which they completed in May of 1967 and presented to Sanders' internal research and development division. By this time, Baer had added Bill Rusch, another Sanders engineer, to the team. Rusch had an understanding of games and brought needed entertainment value to the project. In June, they demonstrated a target-shooting game that used the first light gun controller, and by November of that year, Baer's team finished the world's first ping-pong video game.[1]

And then, in 1972—six years after that September morning and two decades since Loral's fateful request—the world met Baer's invention in a humble machine dubbed the Brown Box, licensed by Magnavox and marketed to the public as the Odyssey.

Unfortunately, it didn't do well. Magnavox's confusing advertising led many people to believe that the Odyssey would only connect to a Magnavox TV, and the console's hefty one hundred dollar price tag (this was 1972, after all) didn't help.

That could have been the end of the story. Baer's invention might have ended up a footnote in the annals of technological history were it not for three other contemporaneous developments. One was an isolated event, one ran on outrageously expensive machines well beyond the public's reach, and one went on to change the world.

• • •

Brookhaven National Laboratory seems an unlikely place for groundbreaking game development. Established by the US Department of Energy (DOE) in 1947 on Long Island, New York, its mission is to support the DOE by conducting cutting-edge research, developing advanced technologies, and educating future scientists. Major discoveries at Brookhaven include synthetic insulin, maglev trains, technetium-99m (a radiotracer used in disease diagnosis), and the Courant-Snyder strong focusing principle, which makes modern particle accelerator design possible. Yet it was here, in 1958, that Willy Higinbotham—nuclear physicist and head of the lab's Instrumentation Division—invented the interactive computer game *Tennis For Two*, ancestor to modern video games.

At the time, Brookhaven held annual visitors days each fall, which saw thousands coming to tour the lab and get a look at the goings on of its scientists. Recognizing that static exhibits just didn't grab attention, Higinbotham tried something different. Using an analog computer hooked to an oscilloscope, he designed a simple two-person tennis game. Players controlled the ball via a knob-and-button arrangement and watched the game on the scope's five-inch screen. Higinbotham was merely trying to relieve boredom. The response was overwhelming: hundreds lined up to play Higinbotham's game. It was so successful that the lab rolled out an upgraded version the following year, which received another enthusiastic reception.

And then it disappeared.

After that second visitors day, the computer and oscilloscope were separated and returned to service in more traditional pursuits. Higinbotham went back to his work on radar displays and nuclear nonproliferation, never bothering to file a patent for his game. And *Tennis For Two*, after existing publicly for less than forty-eight hours, faded into memory. Until a lawsuit in 1985 brought the game back to light,[2] the only people aware of its existence were Brookhaven staff and the visitors lucky enough to have played it—and Ralph Baer was not among them.

Neither was another rather unheralded pioneer—computer scientist and Massachusetts Institute of Technology (MIT) alum Steve Russell. Though he was the first person to implement LISP (the second-oldest high-level programming language still in use today), he's most famous for a more entertaining contribution to the computer world: a little game called *SpaceWar*.

The year was 1961. Steve Russell was a student at MIT and a member of the university's Tech Model Railroad Club—a haven for geeks and model train enthusiasts alike. Russell and his colleagues had just been granted unprecedented access to a DEC PDP-1 (one of Digital Equipment Corporation's first minicomputers) through their professor, Jack Dennis. Small compared to its predecessors, the PDP-1 was a behemoth by today's standards: Though it didn't require floor reinforcement (this from the PDP handbook), it still took up seventeen square feet of space. By comparison, my laptop has a footprint of less than one and has about a half million times more storage capacity. However, it was the most advanced machine of the day, with serious graphics capabilities that begged to be tested. Russell's group was only too eager to take up the challenge. Drawing inspiration from their favorite pulp science fiction novels—E. E. Smith's *Skylark* series—and realizing that the best way to demonstrate a machine's capabilities was to design a program that pushed its limits, had a variable outcome (within the constraints of the program), and engaged its users, they set out to design a game.[3]

The result was *SpaceWar*, a two-person spaceship duel and the world's first interactive computer game. Its premise was simple: two ships hunted each other through space. At center screen, a sun lay in wait, pulling ships inexorably towards it. Stop moving, and you'd gradually be sucked down its gravity well and destroyed. A pass too close would end in similar catastrophe. However, skilled players could use the star's gravity like a slingshot, accelerating around it to take opponents by surprise and torpedo them into oblivion. The only other component was a hyperspace feature, which, if engaged, could get you out of a jam—but also carried with it the risk of instant death. And that was it. From our perspective now, it was beyond primitive. Back then, though, no one had ever seen anything like it, and they couldn't get enough. *SpaceWar* was an underground smash. Russell briefly considered trying to sell the game, but didn't think anyone would pay money for it. Besides, it only ran on the PDP-1—at $120,000, not exactly a consumer machine. So he simply left *SpaceWar* on the development machines, free for anyone to play, copy, or distribute. Russell's team would also give the source code away to anyone who asked for it.

And ask they did. The game went viral, spreading from machine to machine faster than the common cold in a room of preschoolers. Enthusiasts refined and expanded the game, adding their own creative signatures and interpretations. By the mid-'60s, *SpaceWar* existed in hundreds of variations, on virtually every research computer in the country[4]; by the end of the decade, the game had influenced an entire generation of programmers.

One of them was a young man named Nolan Bushnell. He first encountered *SpaceWar* while he was a student at the University of Utah College of Engineering and became instantly hooked, playing the game religiously until he graduated in 1968, committing it to memory. The following year, Bushnell found himself working as a research-design engineer for the northern California–based Ampex Corporation. Though employed full-time, he had an entrepreneurial drive that demanded constant attention. In a move that would ultimately lead to arguably the greatest innovation in entertainment, Bushnell brought his two great loves—engineering and computer games—together, and set about designing a coin-operated version of *SpaceWar.* He developed a working prototype on his own, and then found a partner—Bill Nutting of Nutting Associates—to help manufacture the final game and bring it to market.

The finished game—*Computer Space*, co-created with Ted Dabney—was released by Nutting in late 1971, and immediately went . . . nowhere. Bushnell himself called *Computer Space* a failure, citing the game's complexity as the primary reason: it had a steep learning curve and a set of instructions that ran to book-length. This was a commitment that all but the most dedicated gamers avoided—and in 1971, they were few and far between.

Failure or no, *Computer Space*'s status as the first commercially available coin-operated video game earned it a place in gaming lore. But that was just the tip of the iceberg. The game played a far more significant role as the inspiration for Bushnell's next venture: the founding of Atari and the push to develop what would become the most important game in history.

1972: A *PONG* ODYSSEY

Following *Computer Space*'s lackluster reception, Bushnell and Dabney set out on their own, convinced they could do better than Nutting Associates . . . or at least, no worse. According to Bushnell, "In some ways, it was a blessing to have worked for Nutting. It didn't take very long to figure out I couldn't possibly screw things up more than these guys had. Seeing their mistakes gave me a lot of confidence in my ability to do better on my own."[5]

It was in this spirit that Bushnell and Dabney founded Atari, Inc. Incorporated on June 27, 1972, one of the new company's first acts was to hire University of California, Berkeley, grad and former Ampex employee Al Alcorn. At the time, the waters at Ampex were rocky. There had been some layoffs, and Alcorn's future with the company seemed less than certain. Bushnell's offer looked good, so he jumped ship and signed on as Atari's employee number two. Alcorn barely had time to warm up his chair before Bushnell descended upon him with a project: build a ping-pong video game.[6] It was meant to be a test and was never intended for the public, but Alcorn surprised both Bushnell and Dabney by creating a really fun game. Rather than junk it—Bushnell's original intention when he handed Alcorn the task[7]—they developed it into a coin-op game, installing a prototype in a hole-in-the-wall Sunnyvale, California, bar called Andy Capp's Tavern in September 1972.[8] And thus was *Pong* born.

Two weeks later, the tavern's owner called Alcorn and asked if he could come by and fix the machine as soon as possible, as it had become extremely popular. When he went to Andy Capp's to fix the game, he had no idea what to expect. What he found shocked him: the machine wasn't broken, it was jammed—with quarters. The game's only defect was its undersized coin box.

Pong was a hit.

Three months later, *Pong* clones hit the streets. Atari didn't file patents on the game's underlying technology quick enough to do anything legally, so Bushnell's strategy became a gaming arms race: bring better, more innovative games to market before his competitors. They responded in kind, trying to out-innovate Atari, creating better and better games, pushing the limits of technology. Atari followed suit, and the cycle repeated.

And the rest is history. The runaway popularity of video games led directly to the rise of video arcades a few years later, providing an outlet for both the torrent of games flooding the market and the kids who couldn't get enough of them. In 1975, *Pong* made the leap from the arcade to the family room in Atari's *Home Pong*, filling the void left by the Odyssey and paving the way for Coleco's Telestar,[9] Fairchild Camera and Instrument's Channel F (the first home console to use game cartridges),[10] and in 1977, Atari's own cartridge-driven VCS console—later renamed the 2600. Handheld electronic games had made their debut the year before, with the release of Mattel's *Missile Attack*, followed by *Mattel Auto Race* and *Mattel Electronic Football* in 1977—direct ancestors of modern handhelds and smart devices like the Nintendo 3DS, PlayStation Vita, Apple iPad, and Samsung Galaxy Tab. Two of the most popular video games in history—*Space Invaders* and *Pac-Man*—hit arcades back-to-back in 1978 and 1979. Mattel threw its hat into the home console ring in 1980 with the Intellivision. That same year, Activision—the market's first independent game developer—opened its doors, and Atari brought *Battlezone*—a 3D tank simulator that would later attract the US Army's attention—into the arcades. The early '80s also saw the release of Coleco's ColecoVision and the Atari 5200, the first home consoles to offer arcade-quality graphics. And so it went. Improvement over improvement, innovation after innovation. Follow the line forward, and you'll end up at today's cutting-edge systems: the Xbox One, Wii U, and PlayStation 4—systems so technologically far-removed from their origins as to appear indistinguishable from magic.

It hasn't always been a smooth ride. There were periods of stagnation, crashes, predictions of the medium's demise, the sounding of video gaming's death knell. The path to the top is never straight. For video games, though, it's led consistently up. Today, video games are inextricably linked to our popular culture. The number of people who've grown up with games is greater than those who have not, and almost three-quarters of American households play some form of computer or video game.[11]

And it all began with Ralph Baer's Brown Box. That was the genesis of the video game industry as we know it, the catalyst that made everything else possible.

• • •

Ralph Baer laments the future of science fiction writers. "It's all happening," he says. "What can they write about that will impress anybody? They'll take it for granted, you know?"

We'd been eating lunch together and talking about technological progress, and how quickly science and engineering changed the world. Not long ago, things like touch-screen devices, the Internet, cell phones, and virtual reality existed only in the minds of writers like William Gibson and Neal Stephenson. Now, they're part of our daily existence, no more remarkable than family cars or microwave ovens. Technology is doing things that Baer envisioned thirty or forty years ago: online shopping, wireless game controllers, Microsoft's Kinect sensor. In 1989, he invented a more primitive version of the WiiMote and presented it to companies like Konami and Sega. They passed. At the time, it was just too far out for them.

Baer's heard that a lot. In 1980, he developed a camera that could digitize people's faces and load them into a video game, much like the faces of the pro football players adorning *Madden NFL*'s figures—eight years before *John Madden Football*'s debut. Again, no one bit. "I'm always ten or twenty years ahead," he tells me. "Trying to find somebody who wants to spend money on you is pretty difficult."

That's never stopped him, though. True, at ninety, he no longer moves like a young man, and he freely admits that he doesn't have the stamina he used to. His mind, however, is still ferociously active: he gets sixteen new ideas a day, and he's got a collection of folders three feet thick under his desk—works in progress, just from the last couple of years. And he still gets into his basement workshop every day.

Before I leave, he brings me down there and shows me his latest project: a recreation of Willy Higinbotham's *Tennis For Two*. Then he gives me the tour. Shelf upon shelf of his inventions: *Simon*, modified GI Joe characters (one has a functional metal detector), a toy tank with a working range finder, cars, space ships, fighter planes, a tricycle with a processor that can store sixty seconds of audio. And finally, in a back room, behind closed doors, the video games. Consoles, joysticks, light guns. There's the original Brown Box and its immediate predecessor. Baer describes each item in detail—what components are in them, how they work, when he created them. Then he hands me a controller.

And we play.

In spite of myself, I'm awed. In general, I'm not one to be dazzled by celebrity, but this was different: As a kid, I'd spent vast amounts of time in front of arcade games and my Atari 5200, and now I was playing video pong on one of the first home consoles ever, against the father of video games himself. Heady stuff. We talk while we're playing, and I ask him what the spark was that set him down this path. "Well, it started out with one spot," he tells me, "just to see if I could move something around. It sounds ridiculous, but there was nothing. You have to start somewhere, and the feeling of power you get when you can control something that goes around the screen, when there's never been anything like that before. . . ."

Indeed. That sentiment encapsulates the entire spirit of human invention—the drive to ask questions, identify challenges, and create solutions, and the intense feeling of satisfaction, and, yes, power, when your solution works. And as with all inventions, there are unintended consequences, unexpected results. When Baer made that first single spot move, he could never have known that such a modest beginning was the seed from which a multibillion-dollar industry would blossom. He couldn't have predicted that his invention would spawn an entertainment obsession for billions. And he couldn't have imagined that video gaming, in one of its many forms, would reach its tendrils into virtually every aspect of modern life.

3

LET THE GAMES BEGIN

Competitive Video Gaming and the Birth of the Cyberathlete

The game is maddening. Oil barrels roll endlessly towards you as you desperately climb a steel tower to rescue your girlfriend who's being held captive by a deranged ape. Occasionally, barrels collide, releasing menacing fireballs that race around the screen in an effort to burn you down. It's the ape who's tossing the barrels at you, and if you fail to jump them, hammer them into oblivion, or escape from the rampant fireballs, it's one life gone and back to the beginning. The closer you get to the ape, the faster the barrels come. Reaching the top doesn't get you the girl—far from it. If you do make it, the ape grabs her and carries her off to level two. Complete the second level, and it's off to the third and then the fourth.

It seems deceptively simple. There are only four levels, two controls, and less than a handful of actions. There's a very basic plot, and the objective is clear: climb the tower, reach the ape, get the girl. Except you never get the girl. Even if you clear all four levels, you end up back at the beginning—only faster. And then again. It's distressingly addictive, and one of the hardest games to master—and those few that do still *never get the girl. The most you can hope for is to break the game: The legendary kill screen, achieved after twenty-two levels of play, 117 screens. And then death, in seven seconds.*

• • •

The year is 1982. Mullets are on the rise, video games are exploding, and Kong is king. Against this backdrop, a hero emerges: Billy Mitchell, *Centipede* and *Donkey Kong* high-score record holder, video arcade legend, and future video gamer of the century. Twenty years later, Steve Wiebe—a thirty-seven–year–old out-of-work Boeing engineer and virtually unknown *Donkey Kong* phenom—surfaces to challenge Mitchell's record and claim the *Donkey Kong* crown. And thus was born perhaps the greatest rivalry in competitive video gaming.

The two men are polar opposites: Mitchell is brash, charismatic, overconfident. He has a passion for excellence and is not in the habit of settling for less. Successful in his pursuits and a consummate showman, he is competitive gaming's poster child and superstar. Wiebe, by contrast, is reserved, methodical, socially awkward. He is intelligent and talented—a skilled athlete and musician—but lacks self-confidence. Life, by all accounts, seems to enjoy thrashing him regularly. Some might call him a loser, but he's not a quitter. His burning desire: to be hailed as the best. To rise above all others, to reach the pinnacle and be heralded—and remembered—for his triumph. His target: Billy Mitchell's throne. The stage was set for video gaming's greatest battle, and the hunt for the most contested high score of all time was on.[1]

Like celebrated rivalries throughout history, Wiebe and Mitchell's story is the classic narrative of underdog against champion—a retelling of David and Goliath rendered in 1980's eight-bit glory. Human lore is rife with such accounts, the plot inviolate, the actors morphing into each other like set pieces on some restless and inconstant chessboard: Seabiscuit versus War Admiral, the '69 Miracle Mets versus the Baltimore Orioles, Superbowl XLII's Giants versus Patriots, or 2009's Team Instinct versus Triggers Down.

If I lost you on that last one, don't worry, you're not alone. At the time I began this, I had no idea who they were either. To a substantial and growing fan base, though, their exploits are the stuff of legend: Instinct and Triggers Down are athletic teams who compete under the banner of Major League Gaming (MLG)—one of the most respected organizations in the world of professional electronic sports. The young men who make up the teams' rosters are cyberathletes—pro gamers whose feats of skill on the virtual playing field are no less remarkable than those of their Earth-bound counterparts, and for whom video gaming is a very real and viable way of life.

CYBERATHLETES AND THE LEAGUES WHO LOVE THEM

Ottumwa, Iowa. 1982. Over a cold November weekend, the nation's greatest gamers gathered in this small Midwestern city—self-proclaimed "Video Game Capital of the World"[2]—to take part in the first North American Video Olympics and be written into the annals of popular culture: *Life* magazine, enduring arbiter of social import, was here in pursuit of a story on the video game revolution. It was a celebration of sorts, and the piece de resistance was a *Donkey Kong* match-up between Billy Mitchell and Steven Sanders—both of whom claimed the title of *Kcng* champion.[3] It was the first time two world-class gamers clashed live, head-to-head. By the end of the weekend, Mitchell reigned victorious, preserving his record (which he held for close to twenty years), Sanders was all but finished, and competitive video gaming had been born. However, it would be another fifteen years before the dawn of true professional video gaming and the ascent of the modern cyberathlete.

If you had to mark the end of competitive gaming's infancy, it would be June 27, 1997. That was the day that gamer and former Dallas stockbroker Angel Munoz launched the Cyberathlete Professional League (CPL).[4] Munoz was an aficionado of id Software's first-person shooter *Quake*, and it was after a long deathmatch[5] session, hunting and evading other players within the game, that he became obsessed with the idea of professional gaming. He knew he was taking a risk; this had never been tried in the United States. Until the CPL, hardcore gamers competed against each other through unofficial and haphazard LAN parties: loosely organized events where gamers would converge on a central location, link their consoles or PCs together in a local area network (or LAN), and spend hours joyfully blasting each other into oblivion. Someone would keep score, there'd be running leaderboards updated after each event, but no one was really coordinating things, the only prize was bragging rights, and video gaming certainly wasn't getting the positive exposure Munoz thought it deserved. To bring electronic sports out of the basement and into the public eye, to establish a presence and have professional gaming taken as seriously as any other sport, there had to be a governing body to create standards, set regulations for play and a code of conduct, attract sponsors, organize competitions, ensure media coverage . . . and, yes, put up prize money. "People were having LAN parties," said Munoz, "but they were disorganized. I really think CPL was the first organization to bring corporate sponsorship and standard rules to reinvent the way things were done to present it more as a sport."[6]

By that measure, the CPL was a success. In the decade between 1997 and 2007, the organization obtained sponsorships from companies like Intel, Nvidia, AMD, CompUSA, Hitachi, and Pizza Hut, secured professional gaming's first global television broadcast (on MTV *Overdrive*), paid out more than three million dollars in tournament purses, and distributed an additional two million dollars worth of merchandise. It established league rules and standards of professional behavior—including a drug use and drug testing policy—many of which have been adopted by other leagues or used as a template for their own policies. And it accomplished something that's been central to the popular acceptance of every sport since the birth of the spectator: it created a superstar, in the persona of Jonathan Wendell, the most successful video gamer of all time.

Wendell—known to gamers around the globe as Fatal1ty[7]—was eighteen years old when he entered his first professional event, CPL's FRAG 3, held in Dallas, Texas, in October of 1999. He placed third in *Quake III*, bringing home four thousand dollars; shortly after, he competed in Sweden against the game's twelve best players worldwide, going undefeated in eighteen straight matches and returning stateside as the top-ranked *Quake III* gamer on Earth. A few months later, he was back in Dallas for the CPL World Championships, again playing *Quake III* and again dominating the event—defending his title and walking away this time with a grand prize of forty thousand dollars.

Over the next five years, Fatal1ty went on to win eleven more world finals in five different games—an astounding feat that has yet to be repeated, and which he attributes to both intense practice and regular physical exercise[8]—and earn more than $1.5 million in prize money and sponsorships. He's traveled, all expenses paid, to every continent except Antarctica, as both a competitor and professional gaming advocate, and has launched his own line of branded products—including computer components, gaming gear, and clothing. And he's given back, volunteering for and contributing to organizations like the Make-A-Wish Foundation, Women in Film, the Solid Rock Foundation, and Opportunity Village. Fatal1ty is dedicated, compassionate, intelligent, driven. Like Michael Jordan or Muhammad Ali before him, he's inspired both players and fans, and has become the public face of pro gaming. He is the sport's true ambassador. Certainly, no other athlete has done more to increase the global exposure of professional video gaming or invalidate the stereotype of the thoughtless, unfit, antisocial gamer, while casting his fellow cyberathletes and the sport itself in a positive, valuable, and socially constructive light.

It's a start. But Jonathan Wendell is only one person, and at the time, the effort required to effectively launch pro gaming to a mass audience approached the Herculean. By 2006, he'd also largely stepped out of active competition, taking on a variety of different roles as the sport continued to mature. Adding another wrinkle, in 2008—eleven years after its founding, and before realizing Munoz's grand ambition—Munoz sold the CPL and made his exit from the scene.[9] After a two-year hiatus (to wait out the global recession), the new owners resurrected the league, but it's lost much of its former stature: though still very active, the CPL is no longer at the forefront of professional video gaming. The task of awakening the mainstream to the reality of this burgeoning pursuit would have to fall to someone else.

Enter Sundance DiGiovanni and Mike Sepso, video game enthusiasts, business partners, friends—and both possessed of a similar vision: to build a world-class organization and change the framework of competitive gaming. In 2002, they laid the foundation of what would become the largest and most successful professional gaming league in America: Major League Gaming.

Encouraged in part by the relative success of the CPL, and recognizing a tremendous opportunity to capture the energy and enthusiasm of competitive gamers, DiGiovanni and Sepso sketched out their shared dream: MLG's structure would bear more resemblance to a traditional sports league—say, Major League Baseball or the National Football League—than that of the CPL, and it would be dedicated to delivering the best domestic and international players directly into the homes of gaming fans across the globe—ultimately catching the eyes and interest of nongamers as well, and transforming professional video gaming from a niche activity to a widely accepted spectator sport in the process.

It was a lofty goal—and, it bears repeating, one that was not without risk. But neither was it without precedent. True, the CPL fell short of achieving broad popular support for professional gaming here in the United States. In South Korea, though, World Cyber Games (WCG) was having a very different experience.

Founded in 2000 by International Cyber Marketing CEO Hank Jeong—with financial backing from Samsung Group—WCG traces its origins to that most venerated of athletic institutions, the Olympics. "The format is Olympian in scope—at least the spirit is Olympian," said Michael Arzt, general manager of WCG's US division. "I think that's the spiritual model in terms of how the brand is positioned and the whole idea of global harmony through gaming."[10]

Though for many it may seem odd or even sacrilegious to conflate video games and the Olympics, he has a point. WCG's mission is to lead the development of a healthy cyber culture and promote "the harmony of humankind through e-sports and its embodiment in the Cyber Culture Festival,"[11] stripping away linguistic and cultural barriers and fostering international exchange. It boasts representation from ninety countries, and more than one million gamers take part in WCG events around the world. Individual countries hold regional qualifying competitions and national finals to choose champions who will carry their banners on WCG's global stage. And, like their Olympian forerunners, the anointed battle each other in front of thousands for honor and glory, their triumphs captured in medals of bronze, silver, and gold.

WCG's international reach and its focus on advancing digital culture helped propel it to early success: barely a year after its inception, its first event drew 389,000 gamers from twenty-four countries and featured a total prize pool of six hundred thousand dollars. A year later, in 2002, 450,000 competitors were vying for more than double that amount, $1,300,000. By 2003, the prize money swelled to two million dollars.

There was another factor that played an even more significant role in the tournament's rapid ascent: cultural acceptance. In South Korea (WCG's home base), professional video gaming is a national pastime supported directly by the Korean government. "You'll get fifty thousand people going to an event to watch two guys square off in a video game," said Arzt. "You've got three networks, two of them being the one and two networks for twelve- to thirty-four–year–olds, that are twenty-four–hour gaming networks."[12]

According to Jeong, one thousand Korean kids support themselves strictly by playing video games on the pro circuit; of those, ten make more than one hundred thousand dollars per year.[13] While it's true that salaries for the top pro gamers in the United States can match or even exceed that, and the best teams can score contracts over six figures, the reality for the masses is sobering. In DiGiovanni's estimation, currently only about forty people in this country can make a living as professional gamers. The rest reside in a tenuous state of existence, winning just enough at one event to get to the next, or simply dropping the sport altogether when the financial well runs dry. "I'd like to get it to a hundred," he said. "I think we're a year or two away from that."[14] The bottom line is that pro gaming enjoys a fundamental legitimacy in South Korea that it simply does not have in the United States.

But it will. It's only a matter of time.

In March of 2007, I had coffee with Nina Huntemann in Davis Square—off the Red Line in Somerville, Massachusetts—and we had a long conversation about, among other things, the viability of pro gaming as a spectator sport (remember her; she'll show up again later). Nina's both a good friend and an associate professor in the Department of Communications & Journalism at Suffolk University in Boston, and she's been studying video games for more than a decade. To her mind, the only obstacle to the mainstreaming of pro gaming is cultural approval—and it's an obstacle that will eventually fade.

> All the other things that make spectacle are there: There is glory, there is money—there's money to be won and money to be earned in tickets to watch these tournaments—there's the glamour of the winners, there's the fascination with their skill, the sort of awe with what they're able to do, there's certainly lots of buy-in from hardware companies, software companies. So I think that all of the elements are there for its success. I think that the only thing that's in the way is cultural acceptance, and that'll happen.

It's a thought that's been echoed, with minor variation, by many people across the industry. Often, they compare pro video gaming to other esoteric activities that eventually broke into the mainstream consciousness. When I talked with Joe Minton, CEO of DDM Agents, he had this to say: "You look at poker and say, 'well, ten years ago, the idea of tuning in on TV and watching people play poker, people would think that that was a bit of a joke.' And now it's huge, and there's so many different people doing it."

Matthew Bromberg, MLG's president and CEO from 2006 to 2010, had a similar perspective. "I've always believed that, if millions of people can watch other people play poker, if millions of people can watch people making left turns around a track at high speed, [they can watch pro gaming]."[15]

But would they get it? That's the ultimate question: Is the audience big enough? Are there enough people who actually understand—and would be willing to watch—a professional video game match? Most people I talked to truly believed there were or soon would be. It was simply a matter of numbers and patience: For the first time, those who grew up with video games outnumbered those who didn't. Going forward, the disparity would only increase, and knowledge of video games would shift from the exception to the norm. That's certainly my friend Nina's take on it. "The visual expectation," she said, "the visual language of the game will become a common lexicon, or become a common language that everyone is used to seeing, so that watching all the people on the screen, you'll be able to follow the action because it's a familiar picture, it's a familiar visual experience."

It made sense to me. I've been a gamer most of my life and have engaged in several deathmatches in the past; first-person shooters (the bulk of pro circuit games) were not a foreign language, and I could keep up with the games well enough to hold my own in a casual, friendly bout. *Professional* competitive gaming, though, was terra incognita—and talking about it was only going to get me so far. To really grasp its complete reality, I'd have to experience it for myself.

• • •

When I walked into New Jersey's Meadowlands Expo Center one sunny April morning, I wasn't sure what to expect, but it certainly wasn't this. I'd come to witness MLG's season opener—the first of five stops on a tour that would culminate several months later in the National Championships in Orlando, Florida—and the house was *packed*. Upwards of ten thousand people crammed the floor, looking to cheer on their favorite teams, compete against each other in one of the public open brackets, or even go head-to-head against an MLG pro. Surprisingly, the gamer stereotype was in short supply. Yes, there were a few overweight young men, but you could count them on one hand and have fingers left over. There were people taller, younger, shorter, older. A teen in fishnets. Parents were there—as spectators *and* players—and young women, both attached and unattached (though not lesbians, they pointed out, thus laying to rest another stereotype). There were male teams, female teams, mixed teams. And, of course, there were fresh-faced hopefuls looking to take on the champs.

One of those was a player identified only by his screen name, Halo001. Halo001 wasn't a pro, but he was good. Very good. He was in the event sponsor's booth, playing one-on-one against anyone who dared approach him. I watched in awe as he took down fifteen, maybe twenty in a row—and he'd been at it since before I arrived. He was unstoppable, facing all challengers and sending them out in little balls of pixelated death. I could picture virtual body bags piling up.

And then MLG pro StrongSide entered the hall—laid-back, buoyant, approachable . . . until he took the open seat next to Halo001. Then it was like someone flipped a switch: he meant business. The crowd didn't hush expectantly here: they knew what to expect and were waiting keenly for it to happen.

One match later, they got it. After a slew of victories, Halo001 (never did get his name) was finished, rendered digital Swiss cheese courtesy of StrongSide's expertly wielded battle rifle. From my perspective, it was like bad sex. Or the Kentucky Derby. My first experience watching a pro, and I'd only just gotten into it when it ended. Just like that. Without ceremony or fuss, Halo001's streak was over. He vacated his seat and was subsumed into the gathering crowd. And I got a small taste of what it meant to be a pro. As the competition kicked into high gear, though, and the professional teams began tearing each other apart in four-on-four matches, it became abundantly clear that taking home an event title required much more than individual skill.

My dad used to play pro baseball—minor league, double-A—as shortstop for the Washington Senators. And he was great. He broke records regularly and would steal home so frequently that opposing teams might just as well have inscribed his name on the plate and handed him an invitation to drop by whenever he was in town. From a right-handed swing, he'd make it to first base in 3.3 seconds—as fast as

Mickey Mantle from the left side (Mantle was a switch hitter, and hitting from the left gave him an advantage: the momentum of a left-handed swing takes a batter two or three steps towards first base. Right-handers, whose swing usually takes them away from first, have to make up this discrepancy through sheer speed). He tells a story about a season during which one particular player was hitting exceptionally well. He led the team in both home runs and RBIs, and could rip the ball like other guys went for a summer evening stroll. With him on the roster, the team should have been unbeatable. But they lost. A lot. That is, until he was sidelined due to injury. And though skeptics proclaimed the end of the season—which, on the face of it, seemed a reasonable assessment—for the rest of the team, it was day and night. With him gone, they started winning—because the rest of the guys, though maybe not as hot in the batter's box, knew something that escaped him: to succeed as a team, every player had to be committed.

That the best individual talents don't necessarily make the best team holds true for all sports. And as I was about to find out, it didn't matter whether the field of contest was Yankee Stadium, the Rose Bowl, or Blood Gulch.[16]

It was the final event of the weekend, the championship match that would determine rankings for the next event. MLG underdog Triggers Down had fought their way through the losers bracket, and they now had to defeat the league favorite, Instinct. Amid thousands of screaming fans,[17] I settled in to watch.

At first, I was completely overwhelmed. Following the points of view of four players on one screen is like listening to a quartet of people speaking different languages you've never heard, really fast. It's enough to induce epileptic seizures, or at least motion sickness. However, once my brain adjusted and my eyes got used to what I was seeing, I found it relatively easy to keep up with the action.

Initially, what impressed me was the cool. All the players were focused, concentrated. And utterly unflappable. While the coaches were bouncing around behind them and shouting instructions and encouragement like overhyped speed junkies, the players themselves were *cool.* No golfer putting for birdie from the edge of the green ever displayed more concentration than these kids, eyes locked on the screen, totally aware of their teammates and the opposition, fingers flying like their brains were wired faster than the rest of us. And they were graceful under pressure, displaying sheer feats of skill and stamina that would elude all but the best gamers.

More impressive, though, beyond the level of individual ability, was the way the players worked together—and how that differentiated the two squads. Even to my relatively untrained eye, it was quickly evident that Triggers Down was better. You could see it in the way they played. They all had stand-out moments, but no single player dominated. Each one had his specialty or strength, they carried out their roles expertly, and they all backed each other up. And it showed in the results. While Instinct arguably had better individual players (including team captain Walshy, perhaps the best gamer in the league), Triggers Down was by far the stronger team, and they dominated the finals, losing only two matches out of eight to take the championship match and gain top seed in a turn of events that few expected.

Though Triggers Down was ultimately defeated, finishing fifth overall at the National Championships, personality conflicts and internal shakeups in both teams landed some of its best athletes on Instinct. A couple of years later, I watched the reformed Instinct dominate the MLG National Championships in Providence, Rhode Island, taking home the league title, season trophy, and a check for one hundred thousand dollars.

Now, I'm sure some of you are asking yourselves how it's possible that a team could win a hundred grand at one event, just for beating another team at a video game. After all, how hard can it really be? If you're a gamer, you might even be thinking that you could take on the pros yourself and grab some of that easy money.

Allow me to disabuse you of that notion: You can't. Sorry. As a delusion, that ranks right around believing that, because you're good at slow-pitch softball, you can step into the batter's box and hit a major league pitcher who's throwing heat faster than most people drive. Don't get me wrong, you may be a really good gamer. It's just not going to happen.

An ESPN interviewer once asked Sundance DiGiovanni what he'd tell the average person who thinks he or she might stand a chance against an MLG pro. "Just because you participate in an activity," he said, "doesn't mean you can beat the professionals. You better practice if you want to jump into the arena with these guys, because most of our players practice eight to ten hours a day in preparation for these events.

It's just like any participant or spectator sport. You can . . . be the best among your friends, but then there is going to be a skill gap between you and the professionals."[18] Case in point: Halo001, who, though good against the average event attendee, was annihilated by the MLG pro StrongSide in seconds.

Imagine trying to shoot a moving target the size of softball from one hundred yards away while jumping in the air and avoiding taking a bullet yourself, and you get some idea. That's only one piece of the picture, though. To be the best—to win a national championship—takes more than just hours in the game. A pro gaming team has to master a variety of game types—capture the flag, slayer, king of the hill—on different maps (think of them as virtual arenas). They have to know the locations within those maps of special items that provide competitive advantages and beat the other team to them. And the team coach has to develop strategies for each map—how to effectively clear them, who's responsible for what position, where the team can attack from and retreat to. A typical coach will have binders full of these—like pro football playbooks—and will constantly update them. Together, the coach and the team also review videos of past game matches—both their own, to look for mistakes, and their competitors', to identify strengths and weaknesses. Most importantly, the players have to function and communicate as a team. That means knowing their individual roles within the team, being aware of where everyone is, who needs backup, who has what weapon, and how to improvise when things go wrong.

It no longer sounds easy, does it? And winning a one hundred thousand dollar prize purse doesn't sound like much—especially when you consider that many pro athletes in traditional sports make more per game for doing far less.

Professional gaming has come a long way, though. In 2006, three years after MLG's first event—which drew about 120 people, was funded entirely by DiGiovanni and Sepso, and was an unmitigated disaster—DiGiovanni secured ten million dollars in venture capital (VC) funding. A year later, MLG received an additional twenty-five million dollars, and it's been growing ever since: As of March 2012, DiGiovanni had raised a total of seventy million dollars in VC funding[19]; MLG's operating revenue is now in the neighborhood of twenty million dollars.[20]

Viewership is on the rise as well. Six years after that 120-person audience, MLG's 2012 Spring Championship drew 4.7 *million* unique online viewers (437,000 at once, at its peak) who watched 5.4 million hours of competition video[21]—overshadowing 2012's Sugar Bowl, Rose Bowl, and National Basketball Association All-Star Game.[22] MLG can now be seen around the world in more than two hundred countries—thanks in large part to improvements in streaming video over the Internet. "Streaming has allowed us to take what was an in-the-room experience and broadcast it to millions of people," said DiGiovanni. "If you go back five years ago, that wasn't possible. The costs keep getting reduced and the quality keeps getting better."[23]

Most recently, DiGiovanni announced a partnership between MLG and CBS Interactive. Though this won't bring MLG matches to network television, CBS does own Gamespot—one of the largest gaming websites. This deal gives CBS Interactive the exclusive right to broadcast MLG Pro Circuit events, as well as handle all of MLG's advertising representation.[24] What does this mean? In the words of Daryl Morey, general manager of the NBA's Houston Rockets—expressing a thought that, ten years ago would have been tantamount to heresy—"eSports will someday be the biggest sport in the world."[25]

So, professional gaming is well on its way to becoming the next big spectator sport. The rapid expansion and improvement of streaming technology allows it to reach a global audience. More and more people are growing up with video games—playing them and understanding their language, what they represent. Sponsors are beginning to see the potential, and more money is becoming available for both leagues and players: In the near future, professional cyberathlete will be a viable career for more than a handful of people. Good news for DiGiovanni and Sepso, Jeong and Wendell, and the millions of players and spectators around the world.

But what about the rest of us? Many of the games these cyberathletes are playing—and the fans are watching—are inescapably violent. If professional gaming does go mainstream—and it seems almost inevitable that it will—what are the consequences? I asked Nina Huntemann about this, over that same cup of coffee. She paused for a second before responding.

> Nobody's getting killed in real life in gaming, but we're watching virtual people get killed. And to me it contributes to a culture that accepts violence. Not necessarily a culture that *is* violent—meaning that playing a game, I don't think, translates into acting violently. No research on media effects can prove that, consistently, anywhere. Never, in decades of research. But I think it does contribute further to a culture that accepts violent action. And that may have social . . . I think it does have detrimental social effects. What does it mean when, as a culture, we are entertained by watching two teams "kill" each other through this virtual game? I think that's a cultural question that may lead to, as people have said, desensitization to violence, acceptance of violence as a means to resolve conflict.

A salient point indeed, particularly when you consider the popularity of football—an inherently violent sport in its own right. In fact, you could argue rather convincingly that more people have been injured or killed as a result of football than from video games. And then, there's the stereotypically violent devotees of soccer. But what's interesting about both of those sports is that it's the fans who are typically the perpetrators, not the athletes (ice hockey is, of course, a notable exception). Brawls break out in the stands at football games and soccer matches with disturbing frequency, but the athletes are held to a code of conduct that does not allow for such behavior. Pro video gaming leagues and teams have strict codes of conduct and sportsmanship as well—some even more so, given that many cyberathletes are minors. "This is a potentially positive outcome of the professionalization of gaming," Nina said, "is that as it becomes professionalized, the team members themselves are more likely to adopt a professional attitude." If cyberathletes become role models (as many athletes in traditional sports have become), the way they comport themselves may go a long way towards positively influencing their fans—and, perhaps, society at large.

Obviously, it remains to be seen. However, there's an unpredictable, essentially human element we would do well to remember: Whenever emotions run high, there's an underlying threat of violence—and nothing seems to stir our emotions quite like when a favorite team is being beaten. Fans are fans, whether of the New England Patriots, Manchester United, or Triggers Down; we shouldn't expect them to behave differently.

But neither should we throw up our hands in defeat, faulting the darker side of humanity and bemoaning our inevitable descent into violence. At its root, competition is a unifying force. Carried out on the playing fields of Earth or within the digital rendering of an imagined world, it joins us in support of a champion who fights for our common cause. We rejoice or suffer with them, and in so doing, are drawn together into our humanity. Competition also serves as a lens through which to view society at large, and a mirror from which each of us, as individuals, can reflect on ourselves.

It is human nature to compete, and competition can be brutal and, yes, violent. But it can also be beautiful and inspiring. When an athlete breaks the bounds of the physical and triumphs over the insurmountable, it awakens in us a realization that we, too, are capable of more than we know—forcing us to challenge our preconceptions and push the limits of the possible. And when we witness it together, it unites us all in our collective struggle to slip our Earthly fetters and soar with angels among Elysian fields.

4

ALPHABET SOUP

MMOs, MUDs, and RPGs—*D&D* in the Twenty-First Century

April 18, 2005. The leader of UQS, one of the world's largest corporations, travels to a remote location, attended by a complement of bodyguards and alongside her most trusted lieutenant. As a show of corporate might—and at the suggestion of that same lieutenant—they make the journey side-by-side in their most valuable and imposing vehicles, each worth billions. It's a display befitting an organization that dominates the upper echelons of power and wealth—rarefied terrain inhabited by the barest fraction of the few.

The trip itself is uneventful, but their arrival would prove otherwise. Within UQS, all was not as it seemed: For the last year, a group of mercenaries had been infiltrating the organization, individuals spreading like a virus throughout the corporate structure, rising to positions of trust and influence. Several made it onto the Board of Directors, and one had established himself as the CEO's second in command—the very person traveling with her now. The mercenaries were hired by a rival corporation that wanted revenge against UQS for a shady business deal. The assignment: take down UQS, assassinate the CEO, and bring back her body.

It was a masterful piece of espionage. The mercenaries hiding within UQS were so effective at projecting loyalty to the organization that they achieved unrestricted access to key personnel and the bulk of its finan-

cial and material assets. After a year of planning, they were positioned at critical sites across the entire company and were privy to its most closely held secrets.

It was time to strike.

When the CEO and her entourage arrive at the rendezvous site, the mercenaries pounce. The ambush happens from both outside and within, as several of her personal bodyguards turn on her, augmenting the mercenary force lying in wait. Together, they make short work of her priceless vehicle. She tries to escape, but in an act of betrayal worthy of Brutus himself, her own lieutenant, second in command of UQS and her most trusted confidant, guns her down, collecting her corpse to deliver, as promised, to his client.

The assault on the CEO was only the beginning, though. Once the battle was joined, all mercenary agents concealed within UQS received a single code word—"Nicole"—thus initiating a coordinated attack on every UQS facility they'd managed to penetrate. They looted warehouses and storage hangars, destroyed vehicles and property, stole raw materials, confidential blueprints, and equipment, and emptied the treasury of a lifetime of corporate wealth.

And then it was over. In a single hour, the mercenaries had made off with billions, destroyed billions more in assets, and killed the CEO of one of the most powerful corporations in the world. They didn't just cripple UQS, they brought it to its knees, delivering a blow from which it would never recover.

• • •

This account is a retelling of an actual incident involving the Ubiqua Seraph corporation (or UQS), a mercenary outfit called the Guiding Hand Social Club (GHSC), and an anonymous third party who contracted the GHSC to eliminate UQS's enigmatic leader, a woman known only as Mirial. It's a sordid, though not unfamiliar, story: corporate espionage, wanton destruction, larceny, murder. The GHSC carried out its attack in public, and rather than exercise discretion after the fact, its leader—Istvaan Shogaatsu (his name, like everything else about the event, is common knowledge)—broadcast the group's accomplishment over the Internet. Everyone knew the parties involved, everyone knew what happened. And no one did a thing about it.

How did the GHSC get away with it? And why haven't you heard of this before now? Well, first, it was all perfectly legal. And second, you probably don't play *EVE Online.*

The events of April 18, 2005, took place much as I've described them here, but they occurred within the virtual sci-fi universe of *EVE Online*—a persistent-world, massively multiplayer online role-playing game (MMORPG or MMO)[1] set twenty-one thousand years in the future. CCP Games, *EVE*'s developer, designed the game to be as open as possible, allowing players to engage in a variety of activities similar to other MMOs, including mining, exploration, combat, manufacturing, and trading—as well as more unsavory pursuits like piracy, theft, fraud, extortion, espionage, and murder. Except in very rare circumstances—cases of extreme harassment, for example, like killing a particular player over and over again—CCP permits these endeavors, preferring to let people negotiate *EVE*'s universe for themselves and take responsibility for their own conduct and decisions, good or bad. While this provides gamers with a high degree of flexibility for how they play *EVE*, it can also give rise to Ponzi schemes that bilk players of billions in virtual currency and contract hits that can erase months or years of work in a moment of superbly orchestrated brilliance.[2]

Though many within the game bemoan such dark exploits, it's situations when the lines between the real and game worlds blur that give MMOs complexity and depth. CCP doesn't officially condone such behavior, and they certainly didn't anticipate the creative lengths to which people would go to defraud and betray each other, but they don't condemn it either. To them, it's just a result of providing human beings an open-ended world to explore and exploit however they see fit—the cost of business, if you will, for playing *EVE*. Human nature being what it is, some negative situations arise: People are essentially trusting creatures, and there are those who thrive on manipulating others to their own advantage. You don't have to subscribe to this, but if you want to stay in the game, you do have to accept it—and that's a choice that every player must make for him- or herself. Inevitably, events like the destruction of Ubiqua Seraph drive some players from the game for good. However, even more get drawn to it: Following *PC Gamer*'s publication of the UQS story, *EVE*'s population soared,[3] suggesting that the idea of being able to engineer such subterfuge—that it didn't run afoul of either the game's rules or the developer's code of conduct—outweighed the reality of the event itself. Alex Engel, MMO developer, former CCP volunteer, and *EVE Online* veteran of nearly a decade, shared his thoughts with me one evening before Boston Post Mortem, the monthly meeting of Boston-area game developers. "I love it," he said. "That ruthlessness you cannot find anywhere else in the online game industry."

I mentioned to him that, while many gamers appreciated the skill needed to develop and execute such an audacious plan, there were others who thought it was, in a word, evil. He paused for a second. "Well, yes," he said. "People who thought this thing was evil can easily see it happening to themselves, because living in a trustless world is very alarming. And the whole point of playing online games is to make relationships and social bonds, and work with people. So seeing that trust abused? Man, were people ever pissed."

All is not betrayal and death, though. Those same bonds that made it possible for GHSC agents to establish themselves within Mirial's ranks and wreak such devastation also allow people to create online communities and forge real friendships—not just in *EVE*, but in all MMOs. They go by different names—corporations, clans, guilds—but are founded on the same principle of bringing people together around a shared interest and working towards a common goal. Devotees of Blizzard Entertainment's fantasy-themed MMO *World of Warcraft (WoW)*—one of the genre's most popular and successful titles—are no strangers to this. Though you can go through the game solo, many players choose to form or join alliances with other like-minded players, called guilds. Guilds usually include a wide range of people—high school and college students, club bouncers, couples and families, orthopedic surgeons, university professors, active-duty military personnel—who can hail from anywhere in the world. Members meet up for quests, raids on other guilds, or simply to explore the world together. Typically, they're hooked up through some form of Internet voice chat (Ventrilo is one of the most popular), and they talk to each other while playing—frequently about things completely unrelated to the game: they talk about work, about their kids, family life, anything you can imagine, from the mundane to the exceptional. And therein lies the secret to *World of Warcraft*'s appeal: for many, it's not the stunning visuals, engrossing game play, or complicated tasks that draw them, but the interaction with other people, and the relationships that arise as a result. The same can be said for most MMOs. It's one of the aspects of *EVE* that keeps Engel playing. "I have people that I call friends—and count them as friends—that I've known now for eight, nine years," he told me. "You may spend anywhere from thirty minutes to five hours a day talking to these people [in-game]. I mean, they're your real-world friends just as much as talking to your friends on the phone."

It's a social dynamic I'm very familiar with, not through games like *EVE* or *WoW* (neither of which I've played myself, for fear of becoming irretrievably lost in their virtual realms), but through their low-tech, die-and-paper forerunner, common ancestor of every modern MMO in the cybersphere. I'm speaking, of course, of the original role-playing game (RPG): *Dungeons & Dragons*.

Yes, I played *D&D*. Loved it, actually. Most weekends, you could find me and my friends holed up in the living room, hunting treasure and locked in mortal combat with orcs, goblins, trolls, and dragons. There it is: confirmation of my status as geek. But it wasn't simply about escapist fantasy. We, like *D&D* geeks everywhere, were caught up in the teamwork, the camaraderie, the—dare I say it—inherently *social* nature of the game.

It may go against the cultural grain to call *D&D* social, but anyone who's played it is well aware of this fundamental truth. Sure, you *can* run the game with only two people, but the best campaigns are group experiences. We regularly played in a company of eight to ten, and together embarked on epic journeys and carried out heroic deeds, allowing our imaginations free rein. More often than not, though, we'd get sidetracked by banter and conversation—catching up with each other, discussing events of the week, or just enjoying the simple pleasure of being in a room with good friends.

And we weren't alone. Created by Gary Gygax and Dave Arneson and published by Tactical Studies Rules, Inc. (TSR), in 1974, *Dungeons & Dragons* met with steady and growing success: Though the first run of one thousand copies took them almost a year to sell, the second and third printings, in 1975 and '76 (two thousand and three thousand copies, respectively), sold out in five months.[4] By 1976, TSR had three hundred thousand dollars in revenues; by 1982, that number had grown tenfold, to more than three million dollars.[5] When my friends and I encountered the game in the mid-'80s, we did so in good company: *D&*

D had, by then, attracted four to five million others. Today, it's been played by upwards of thirty million people, all around the world.

There's more to RPGs than *Dungeons & Dragons*, though. *D&D* was but the first. Its success lead to a myriad of others, set in time periods, locations, and galaxies of wondrous variety: there's superhero-themed *Champions*, the James Bond-esque agents-and-espionage RPGs *Top Secret* and *Spycraft*, fantasy games *RuneQuest* and *Call of Cthulhu*, *Vampire: The Masquerade*, *Warhammer Fantasy Battles*, and, somewhat predictably (and rather unoriginally named), the *Star Wars Roleplaying Game*. With so many different RPGs to choose from, and so many people playing (many of them self-professed computer geeks: like the imagination, geekdom knows no bounds), it was only a matter of time before they made the leap to the computer.

It happened in England in 1980, when two Essex University students—Roy Trubshaw and Richard Bartle—created what has become the oldest virtual world in existence. They called it *Multi-User Dungeon* (*MUD*) in honor of one of Trubshaw's favorite PC text adventures, *Dungeon* (a variant of *Zork* adapted for the programming language FORTRAN). Released in the late '70s, *Zork* was one of the first interactive fiction computer games[6]—itself based on an even earlier game, Don Woods' *Adventure*.[7] In *Zork*, you take on the role of a nameless adventurer seeking fortune and glory among the ruins of a vast and ancient underground empire. The game is entirely text-based: you interact with the world via simple commands ("go right," "pick up sword," "attack troll," etc.) and are presented with the results of your actions through descriptive narration. *Zork* has been consistently hailed as one of the best in its genre, especially for the quality and depth of its story. As a teenager, I vividly remember spending hours working through the game on an old Apple IIe, becoming thoroughly engrossed in its world. The story was compelling, the puzzles challenging but not impossible. It contained much of what I loved about *D&D*. What it lacked, though, was true interaction, the shared experience with other players.

That's what Trubshaw and Bartle set out to do with *MUD*: build an online adventure game that could be played—and shared—by multiple users over the Internet. At first, most visitors to *MUD* were fellow Essex University students—and night after night, the server was packed. Before long, however, the university opened up access to the general public, and the game exploded. *MUD* drew an improbable mix of players from around the world who, through a combination of the game's role-playing dynamic and digital anonymity, could be whoever they wanted. Trubshaw and Bartle had inadvertently created the world's first digital community: a group of players who bonded predominantly inside the context of a virtual world.[8]

Once virtual ground had been broken, multiuser dungeons (MUDs) took off. The '80s and '90s saw an explosion in development, with people creating MUDs that stretched beyond their fantasy-themed origins (though there were new ones in that field as well). Alongside their fantastical brethren, there were sci-fi MUDs, MUDs based on books and movies, historical MUDs, MUDs centered around player versus player combat (PK MUDs), roleplaying MUDs, talkers (essentially online chat environments), MUDs designed specifically for educational purposes, and the direct ancestor of MMOs, graphical MUDs, which employed a combination of text and graphics to present their worlds.

As expansive and interactive as these multiuser worlds were, they lacked one crucial element that's become a hallmark of every MMO or virtual world in existence today: persistence—the ability of the game world to carry on in the absence of any particular player. But that, too, was inevitable. All it required was time and inspiration, which, in this case, took the form of England-born, Texas-raised game designer Richard Garriott—known more famously to gamers the world over as Lord British.[9]

Garriott got his start with computers at Clear Creek High School, convincing school administrators to allow him to develop a self-directed programming course (during which time he created fantasy computer games on the school's teletype machines). In the summer of 1979, he began working at ComputerLand and came face-to-face with Apple's newest machine, the Apple II—and more importantly, with a game called *Escape*. *Escape* players navigated through a maze presented in the first person, and it inspired Garriott to code his first commercial hit, a first-person perspective dungeon adventure called *Akalabeth*. Built for fun, ComputerLand's owner convinced him to market it, which, in 1980, he did, selling thirty thousand copies at five dollars each. Later that year, he created the first game in his renowned *Ultima* series—a simple single-player fantasy RPG that I spent many hours fighting through and exploring. Over the next two decades, Garriott produced twenty more titles in the series, as well as a handful of other games, driving the industry forward on a number of occasions. In 1984, he released *Ultima II*, which became the first RPG to incorporate consequences and long-term effects. For example, players could no longer wantonly kill everything in sight; they had to consider the outcomes and worldwide implications of their actions. This represented a major leap forward and afforded the genre a depth it didn't have previously. All of a sudden, you had to make choices when dealing with objects and characters within the game. Say you encountered a traveler while wandering down a path during an *Ultima II* session. You had a choice as to how to interact with him—ignore, help, rob, or attack, for example. Your decision might have serious consequences later in the game: if he had a key piece of information that could assist in overcoming a future obstacle and you killed him, you might have a harder time progressing past that obstacle. If you helped him, perhaps he'd reveal something of importance, or grant you a gift of thanks. If you ignored or tried to rob him, you might risk being attacked yourself. The incorporation of long-term consequences meant that gamers had to pay closer attention to the game world and really think about how they approached it, which elevated the standard RPG from a simple hack-and-slash quest to something resembling a true, interactive story—one where the gamer didn't merely respond to the presentation of events but actually had agency in determining their course.

Arguably, though, Garriott's greatest contribution to gaming came in 1997 with the unveiling of *Ultima Online*. Not only did this mark the transformation of the *Ultima* series from single-player RPG to MMO, it was the first commercially successful, widespread *persistent* world,[10] and it paved the way for every MMO that followed. Persistence is crucial to an MMO because multiple people can play it at any one time, and the game world has to be consistent for all of them. If, for example, you destroy a tower or build a bridge at a certain location, everyone else within that MMO should see—and be impacted by—the change you've made to the environment, even when you quit your current game session. Also, if there's a specific item of great power or value that generates only once on any particular game server and you happen to claim it, then it's gone, unavailable to anyone else on that server (this has implications that we'll discuss in a bit).[11] It also means, though, that the game continues even if you're not present. In other words, the world persists, regardless of an individual player's absence.

This kind of persistence is what allows long-term social interaction to really take place, either between individual players or within larger groups like *WoW*'s guilds or *EVE*'s corporations. For instance, two players can meet, spend time together, and then go their separate ways, perhaps exiting the game. When they meet again, the same amount of time will have passed in-game for both of them, and the same events will have occurred—including any adventures each player may have undertaken. In the case of a guild, a persistent world means that all members don't have to be at hand for the guild to carry out tasks, complete quests, or otherwise advance in the game. When absent members re-enter, they'll see any progress their guild has made. Of course, the entire point of joining a guild is to interact and socialize with other people, and so players often arrange times to play together. This is especially true when a guild is planning a large-scale raid or quest, and needs most, if not all, of its members to succeed. For these events, guild leaders reach out to their complete membership, coordinating schedules of people potentially from all over the world to make certain that they have sufficient numbers to prevail.

Victory requires more than mere numbers, though. It takes the right mix of characters of adequate skill, strength, and experience—and the only way to gain those is by playing. The onus, then, falls to each individual member to ensure that his or her character is up to snuff, and this takes time. Suffolk University Associate Professor Nina Huntemann has spoken to many avid *WoW* gamers about this, and she's become fascinated by what she calls the blurring between leisure and labor.

> The amount of time that is put into developing your character and "working" at increasing your character's level starts to look a lot like labor. And when I talk to gamers, they talk about it like it's work. They have to log a certain amount of hours every day in order to make sure that they keep up with their peers in their guild, and they put in enough hours to acquire the kinds of experience and artifacts and objects in the world to be more and more powerful.

For young adult gamers who are still living at home and have comparatively few responsibilities, this isn't a problem; they can grind through the game, carrying out the tasks required to level-up their characters.

But for those who are older and who may have to balance gaming with work and family time—what many would call "real life"—this can be a struggle. And there's no shortcut: everyone starts the game at level one, and the only way to increase your level (and therefore your strength, experience, and ability) is to put in the hours.

Unless you can find someone to do it for you—someone, maybe, who's paid to level-up less powerful characters. Which is exactly what happens. It's known as power leveling, and an entire business has grown up around it. Essentially, it works like this: A company hires a workforce of skilled gamers and pays them hourly to take customers' characters and play them to desired levels (specified by the customers). The company then sells the characters back to their customers at a substantial profit. "The price varies on supply and demand—very basic economics," Engel told me. "Game accounts in general, across games, will range from fifty dollars for a relatively low-level account to, the most I've ever seen an account go for is five thousand dollars. I've heard rumors of higher prices, but I've never actually seen them."

In Engel's experience, accounts sell for, on average, in the low to medium hundreds of dollars. According to one website I visited, to power level a newly minted *WoW* character to level 90 (the current cap) would run you $475.02. This, for an intangible item that exists only within the context of a video game and has no intrinsic value outside of it. You can't do anything with it except wander around a virtual world collecting virtual items and fighting virtual monsters. So why would you pay for this? Well, it turns out that to make that 89-level jump takes about 190 hours—at forty hours per week, that's just shy of five full weeks of work.[12] Sure, you could level it yourself if you had that much spare time on your hands. I don't know anyone who does, though, and I suspect that those who do have it would much rather spend the time engaged in pursuits other than grinding through *WoW*. Like just about anything else, it all comes down to time versus money. As Engel said,

> It helps to think of yourself as having a billable rate where every hour of your life is worth "x" dollars. A lot of players believe that the most fun occurs at what's called the elder game or the end game, and that's when you're at the maximum level and you can participate in the hardest content. So they feel like the content prior to that is trivial, and, really, that it just gates them from going from the non-fun part to the "fun" part. So instead of spending . . . their time

> getting to that point, they're willing to make a value judgment, saying, "well, instead of me investing three- or four-hundred hours of my time, I'm going to spend three hundred to four hundred dollars and purchase an account where someone's already invested that time."

EVE Online provides perhaps the best example of this. Unlike *World of Warcraft*, where the time it takes to level-up is affected by how quickly and frequently you carry out actions—defeating monsters, completing quests, things of that nature—in *EVE*, your progress is based on skill points, which you receive over a calculable, measurable amount of real-world time. There's a minimum commitment required to fly the best, most powerful ships, and unless your account puts in the time, you'll never get to that point in the game. And it's a significant chunk: around four years, according to Engel. For someone in a rush to get to the end game, or who simply wants to avoid being repeatedly vaporized by more advanced players, paying even a few thousand dollars for a top-level account starts to make a whole lot of sense.

There's another service that time-strapped MMO enthusiasts can outsource to a professional, and it operates on the same principle of time versus money. Depending on the game, it's known variously as gold farming, platinum farming, energy credits, influence farming . . . the list goes on. It's typically referred to as gold farming, but regardless of name, it amounts to the same thing: using real money to purchase virtual currency.

Every game has its own monetary system, and the only way to buy items within a particular game—weapons, armor, raw materials, starship upgrades, fuel—is to use the coin of the realm. In general, there are only a few ways to gain currency: trading or selling items between players, finishing tasks or quests, being hired by other gamers for specific jobs, or looting the bodies of vanquished foes. The common factor here is, of course, time: you either spend it to acquire money directly or to find items that you can then exchange for money.

Wouldn't it be easier if you could just cut to the chase and buy in-game currency with cold, hard cash? That's where gold farming comes in. Very simply, gold farming is the process of playing through an MMO and collecting as much of the game's currency as possible, then offering it for sale in the real-world marketplace. Like power leveling, gold farming companies employ gamers skilled at the most popular MMOs to harvest game servers of virtual money. A customer looking for a specific currency—say, *EVE Online*'s ISKs or gold pieces for *WoW*—can check out a gold farming outfit's website, review the current exchange rate,[13] and, if satisfied, place an order for however much s/he needs. If the company has the requested amount on-hand, they simply deliver it to the client electronically. Otherwise, one of their staff plays the relevant game until s/he's accrued enough, and then they make the delivery.

Gold farming got its start in 1997, shortly after the release of *Ultima Online*. The game's near instant popularity provided a fertile market for the exchange of virtual goods, and its in-game monetary system provided an opportunity for enterprising players to turn a profit. All they needed was an efficient means to do so, and they found that in eBay. Established just two years prior, eBay provided the ideal platform for players in possession of unwanted items to sell them to others.[14] And that's exactly what happened. Players discovered that they could list anything—gold pieces, armor, weapons, supplies—and other players would engage in contests to outbid each other, just as if they were bidding on an autographed baseball or rare movie poster. As with traditional eBay auctions, the more exceptional or desired an item, the higher a price it fetched: while there might be a limited market for common gear, something that shows up only once or twice on a given server, like a sword that does +30 damage, could sell for hundreds of dollars.

In 1999, Sony Online Entertainment's MMO *EverQuest* expanded the potential market to five hundred thousand people, greatly increasing the opportunity. At this point, gold farming was still a cottage industry, existing predominantly as a player-to-player exchange. Dedicated gold farming organizations hadn't yet broken into the business, and individuals could still make a decent profit. In 2003, journalist and gamer Julian Dibbell made more than eleven thousand dollars (that's real-world, take-it-to-the-bank US currency) selling gold in *Ultima Online*.[15] Not a livable wage, by any means, but not too shabby, considering that what he was selling didn't actually exist—a point not lost on Dibbell himself, who, at the end of his year-long experiment, tried unsuccessfully to convince the IRS that virtual items had real-world value.[16]

By this time, gold farming had come to the attention of a few wealthy entrepreneurs—most notably Brock Pierce, who founded

Internet Gaming Entertainment in 2001 to sell gold in *EverQuest*, and John Yantis who, in 2002, was netting around a million dollars in annual income, also selling *EverQuest* gold[17]—but it was Blizzard's introduction of *WoW* in 2004 that kicked the practice into high gear and brought it serious notice. "It wasn't really until *World of Warcraft* that a true global market came up for it," Alex Engel told me.

> *WoW* was the first time that you had literally millions of potential customers. Instead of *EverQuest*, where we had five hundred thousand subscribers, and of that five hundred thousand, 5 percent are willing to purchase stuff from us—so twenty-five thousand potential customers—now *WoW* has three million customers [in 2004], and 5 percent of the three million customers are willing to buy in-game gold. Well . . . now we have 150,000 customers. So there was really an economy of scale, and also the sheer desirability of the game and its incredible popularity drove prices up and profit margins actually became quite substantial.

Substantial, indeed. By 2006, the global virtual goods market was worth an estimated $880 million dollars a year.[18] By 2008, that number had climbed to around two billion dollars.[19] And by 2010, it had jumped to more than ten billion dollars annually, in hard currency.[20]

Gold farming is a serious business, by any measure. Its future, though, is anything but certain. In the first place, though the gold farming market is built on virtual goods, it is still bound by the laws of real-world economics: whenever an MMO publisher adds more currency into a game (for example, by updating the game with new rewards), it creates inflation and devalues it against, say, the US dollar, thus making it less profitable to harvest. And it's not that MMOs will disappear. Yes, the games themselves will fluctuate, some going offline and new ones replacing them, but the genre as a whole is all but guaranteed to survive. Just as the real world avoided apocalypse in 2012, so too did the virtual: As of year's end, four hundred million people around the world were playing some type of MMO.[21] To put that into perspective, that's about one hundred million more than the entire population of the United States. *WoW* alone boasts 9.6 million subscribers, and *EVE Online*'s population is larger by a third than that of its developer's home country.[22]

No, the most direct threat to gold farming comes from the game companies. Of that ten billion dollars or so, they see exactly none of it. That's a lot of money, and developers are understandably upset with the idea of someone else making such healthy profits from the fruits of their labor—either through gold farming or power leveling. Both of these—and any other transfer of virtual goods for cash—fall into the category of real-money transactions (RMT) and are expressly forbidden under the end-user license agreements of every Western MMO on the market. In practical terms, that means that any player caught engaging in such transactions, either as a buyer or a seller, can have his or her account permanently banned. Publishers have also managed to exert influence in the area of online auctions: Pressure from a variety of sources—game developers included—forced eBay to shut down all auctions of virtual items and update its terms of service to forbid them going forward. "You can no longer sell either accounts or in-game items for real-world money on eBay," Engel told me. "That's a violation of eBay's terms of service, in addition to being a violation of the code of conduct and terms of service of whatever game you're playing."

There are, of course, ways to legitimately trade items between players: They can meet face-to-face within a game, exchange goods through email (using an in-game mail system), or list items through in-game auction houses and sell them for game currency. These are all perfectly acceptable, provided that no real money changes hands. Of course, acquiring the items to sell or the in-world currency to buy them still takes time, which is why the RMT market took off in the first place. Developers recognize this, and they're exploring different ways to address the issue. At the end of 2007, game publishers Sony Online Entertainment, Funcom, and Acclaim began working with a startup called Live Gamer to provide players an officially sanctioned way to conduct RMTs through their games. And a 2009 update to *WoW* included a feature allowing gamers to teleport directly to battles, drastically reducing search time and helping players advance faster.

Still, gold farmers are adaptable, and as long as there are people who have more money than time, they'll find ways around any obstacles the developers can dream up. In Engel's words, it's a constant arms race. When eBay outlawed auctions of virtual goods, websites cropped up overseas to fill the void.[23] When game companies began recognizing the play patterns of gold farmers (running the same route through a game over and over, repeatedly killing the same monster or group of monsters and waiting for the loot drop, etc.), the farmers started hacking into active player accounts, stripping characters of all their goods and gold, and then offering them for sale.[24] If there's a market for their service, gold farmers will devise a way to meet it.

Game companies don't have the time or resources to chase down every terms-of-service violation or track and ban every suspected gold farmer in cyberspace. But they may not need to. There's an even stronger force coming to bear against them that may prove more effective: the gamers themselves.

For most enthusiasts, an MMO is more than just a game, it's an expression of real life. They take their chosen game very seriously, putting the work and time in to create and develop a character, playing the game as intended by the designers. And they look unfavorably on anyone perceived of as buying into the game or paying for advantage. Engel explained it to me. "There's a lot of social anger and social discontent towards people buying in-game currency," he said. "If you do buy gold or you buy credits or you buy an account, you better not tell anyone, because if you're in your own clan or guild, you'll be probably kicked out, you'll be reported to the GMs [Game Masters], you'll be chastised by all your friends, and they may stop playing with you. I mean, that's how nasty it is."

Nina Huntemann had a similar take. "There're plenty of people on the boards online that argue that it's cheating. You play by the rules, literally. And you don't hack it, you don't change it, you don't hire somebody else to play for you. And like anything else, it's cheating, and get off this game."

There will always be those who espouse the philosophy of, as Nina put it, "by any means possible," whether it be buying, cheating, or stealing your way in. But for someone who takes a larger view, who gets into an MMO for the social interaction, to develop personal connections around a shared passion, the kind of vitriol directed at gold farmers and the spectre of being ostracized or permanently banished from the community will most likely prove more of a deterrent than the threat of an account ban ever could. It's one thing to lose a game character, after all; it's something else entirely to lose your friends.

And that's really what we're talking about here. For people like Engel, being a gamer isn't just a hobby or even an occupation. A gamer is who he is—not in total, but certainly in large part. His identity, and those of his friends and coworkers, is intimately connected to it—and that ties them to each other. Video game communities are tightly knit, bonded by the collective love of a singular pursuit. "We have this great underground culture that's no longer underground," Engel told me. "You can walk into a group of people and feel a kinship with them that you don't feel when you walk into a random crowd. And that kinship is a love for video games and a love for gaming culture."

I related a story to him, something I heard from a twenty-something gamer at a conference in Boston. It was several years earlier, and this young man's life had bottomed out. He was physically ill, emotionally damaged, depressed, and near suicide—not the picture of the man I was listening to. "Gamers," he said, choking back what could only have been tears of gratitude, "saved my life."

It was a tale Engel had heard before. "In any community that you're a part of," he said, "you'll find people that are willing to support you and bring you up. I hear about it all the time." He took a moment's pause.

> You know, in many ways, we're reluctant to become friends with people because we don't have anything in common with them. It can be really tough to just walk up to someone and say, "I'm whoever I am and you're whoever you are, and I want you to be my friend." It's really tough to do that. But when you have something in common, whether you're going to the gym, whether you're playing video games, or whether you're reading the same book on the subway, it gives you a bridge, and you can cross over that bridge to the other person and make friends. Mutual shared experiences, that's their bridge. And they can use that to bridge themselves from stranger to friend. And from there it's only a short leap to helping someone out.

This, finally, is what we're all searching for from communities we join or those we build with kindred souls: friendship, trust, support . . . even love. We want to be who we are and be accepted for that. And we want to know, without question, that when the chips are down, someone has our back—whether we need a shoulder to cry on, a hand to lift us up, or a Caldari Interceptor to rescue us from ambush among virtual stars.

5

NO CONSOLE REQUIRED

Casual Games (or, Gaming for the Rest of Us)

It's been a long day; you've been on your feet for hours. The kids are finally asleep, and the TV's on in the next room. At last, some time to yourself. You wake up your smartphone and page over to your apps for a quick hit of digital R&R. Your finger hovers: Angry Birds? Solitaire? Mahjong Towers?

Bejeweled. *Yeah, that's what you need. A little match three to take the edge off.*

As the game starts, you can already feel the day's stresses recede. Another moment and you're lost in it, lining up shimmering jewels and watching them explode in a satisfying mix of color and sound. The rhythm of play draws you in, pushing the outside world into the background as group after group of jewels erupt in polychromatic flashes. Your concentration is total: For the next five or ten minutes, the game is your reality.

By the second screen, you've achieved near Zen-like focus; by the third, you've lost all sense of time. When you do come out of it and return to the here and now, you find, to your surprise, that half an hour has passed in an eye-blink. You yawn and stretch: It's getting late, you're a little tired. You've earned three new badges, hit a new high score . . . and you're less edgy than when you started. You should probably get to bed. But there is that new achievement you've unlocked. Maybe just one more game. . . .

• • •

Still with me? Good. We've covered a lot of ground so far, but we've got some distance yet to travel. Now might be a good time to break into those supplies you packed at the beginning of our trip, and check your flashlight batteries: the road from here winds and meanders, and is often cast in shadow.

Okay, let's get moving. Brace yourself, though, because this is where the story gets personal. If you don't acknowledge it already, this is where I convince you that you're a gamer.

Don't believe me? Alright. Question, though: do you own a smartphone? Feature phone? Tablet? Handheld or iPod? Do you visit Facebook or some other social media site on your laptop or desktop? You know those little games you play on the subway, the bus, in the evenings at home or between office meetings when nobody's looking? It's only for a few minutes. You're not wasting time. Just killing it when there's nothing pressing, during downtime between responsibilities.

It's okay, you're not alone. Everybody does it: your colleagues, coworkers, friends and relations . . . even your boss. In fact, fully 40 percent of the entire US population plays video games through a casual website or social network like Facebook. That's 145 million people in this country alone[1]—a decent count in its own right. Add the rest of the world to this rather provincial view and the numbers get truly staggering: 235 million people play social games through Facebook, another four hundred million get their game fixes from other websites, and close to four hundred million more play games directly on their mobile devices—smartphones, tablets, mini-tabs, and iPods.[2] Give it a second, it'll hit you. That's right. More than one *billion* people. Nearly one-sixth the population of the *planet*.

See? You really aren't alone. Not only that, you're actually part of the largest and fastest-growing segment of the video game industry: more people play casual games than games on Xboxes and PlayStations combined. Even those so-called hardcore gamers enjoy casual games—so much so that it's affecting their regular gaming: a third of them spend less time hooked up to their consoles since they began playing social games on networking sites.[3] And the number of casual gamers is building. Driven largely by an increase in smartphone ownership, eMarketer estimates that by 2016, mobile gamers will make up more than half the entire population of the United States.[4]

Let me back up for a minute. I've thrown a few labels out there—casual game, social game, mobile game—and you may have noticed that I'm using them somewhat interchangeably. So what's the difference between them? And what, exactly, is a casual game?

Well, it depends who you ask—and I've asked a lot. Here's the problem: things change—technology, players' behaviors and expectations, the gaming audience itself—and developers build games that reflect these changes, taking advantage of a new gaming platform, for example, or incorporating new functionality or gameplay elements into their games. This dynamic character makes coming up with a specific definition of casual games a bit like throwing a bull's-eye on a moving dartboard while simultaneously executing a spinning leap through the air. During the summer of 2012, I spent three days in Seattle at the annual Casual Connect Conference, and I saw more flavors of casual games than I thought possible, in varying degrees of complexity: first-person shooters, puzzle and word games, role-playing games, world builders, racing sims, virtual zoos, hidden-object games, casino games, bubble poppers, interactive story games, match three—something for every age, personality, and interest. I could go on, but you get the idea: There were *a lot* of games on display. And though they differed in mechanics, appearance, and objectives, they all shared three characteristics—and after speaking to several people at the conference, I found that this was a broadly accurate and generally accepted description, if not a hard-and-fast definition: Casual games are easy to learn yet hard to master, can be enjoyed in small slices of time (typically between five and twenty minutes), and can be played almost anywhere. Matt Hulett, then president of Seattle's GameHouse,[5] put it best. "They're snacks," he said. "Most of our Facebook gamers, they play, on average, eight times a day, for one to two minutes at a time. By definition, that's snacking." So, while hardcore console games like *Halo* and *Call of Duty* are full meals, casual games are lighter fare. Of course, that doesn't necessarily make them less compelling. Yes, casual games are designed to be consumed in small doses, but the best can capture you for an hour (or more). I know whereof I speak: my personal addiction is PopCap's *Bejeweled Blitz*. A single game takes one minute to complete. I don't think I've ever played less than ten, and that's on a slow night. Usually I blow through fifty or sixty without even thinking.

As a category, casual games is exceptionally broad and encompasses virtually all game genres, which adds to the difficulty in defining it. Under its umbrella, though, there are four main subgroups that we can dump all casual games into, based primarily on how they're experienced or delivered: casual computer/PC games, downloadable try-and-buy, social games, and mobile games.

Those of you who are older may remember the games that came bundled with Microsoft Windows in the early 1990s: *Solitaire*, *Minesweeper*, and *FreeCell* among them. Designed to help ease the anxiety of users new to Windows—as well as to test components of the operating system—they were many people's earliest exposure to casual games. The most famous casual PC game—and arguably the first—wasn't bundled with any operating system, though versions were eventually released for most of them: *Tetris*. Designed by Soviet computer engineer Alexey Pajitnov in 1984, the game made it to our shores in 1986 and became an instant hit. For those of you who don't know *Tetris* (perhaps even more rare than those who haven't seen *Star Wars*, but they are out there), allow me to elaborate. Colored blocks in a variety of shapes drop endlessly from the top of the screen and collect in an ever-growing stack on the bottom. The player's job is to arrange the blocks such that they make complete rows, which then disappear, reducing the height of the stack by the number of rows eliminated. There's a catch, though: the blocks don't fall at a constant speed, they accelerate: the longer you play, the faster they fall. Eventually, no matter how good you are, it becomes impossible to keep pace with them. The block tower rises to the top of the screen and the game ends. That's it. Ultimately, the game is unwinable. As popular today as when it emerged from eastern Europe, *Tetris* exemplifies the ideals of a casual game: easy to get into, tough (if not impossible) to master, and ridiculously hard to quit.

The rise of the Internet and the demonstrated success (particularly id Software's *Doom*, in 1993) of the shareware model of game distribution—which allows users to download and play a two- or three-level trial of a game before purchasing the full version—ushered in a new phase of casual gaming: downloadable try-and-buy. Following id's example, casual game developers started offering "lite" or time-limited versions of games for people to download to their PCs, try out, and then, hopefully, buy the full versions. Often, these were simple puzzle, action, and word games targeted at a nongaming audience, predominately middle-aged women.

If you spend any time at all on Facebook, you've been exposed to social games. They're everywhere: *FarmVille*, *The Sims Social*, *Bejeweled Blitz*, *Bubble Witch Saga*, *Mafia Wars*, *Zynga Poker*, *Dragon City*, *Words With Friends* . . . you're either playing one or have been invited to play one—perhaps repeatedly. Social games usually require a player to be online, run embedded within the network, and incorporate their social nature directly into the gameplay. If you're playing *FarmVille* on Facebook, for example, you can send invitations to your Facebook friends directly through the game. The more who join you, the faster you can make progress in the game. You can also give and receive gifts or assistance to or from your Facebook friends who are already playing. There are around seventy-eight million social gamers in the United States today, and though the games are targeted at a more diverse audience than PC downloadable games, forty-year-old females still represent the average.[6]

Of course, you can't talk about casual gaming these days without bringing up mobile—and with good reason. If casual gaming is growing faster than any other segment of the industry (and it is), it largely has the mobile sector to thank for it.

Mobile gaming is exactly what its name suggests: gaming on a mobile device—smartphone, tablet (full-size or mini), or handheld, like an iPod Touch. It's the youngest component of casual, inaugurated in 2007 shortly after the introduction of Apple's iPhone, but it's also shown the most rapid growth: As of March 2012, the number of American mobile gamers passed one hundred million—a 35 percent jump from 2011[7]—and it seems all but unstoppable.

There are three aspects of mobile gaming that spike developers' adrenaline. First, games built for mobile devices can use the features of those devices—GPS, camera, microphone, position sensor, and video—to enhance gameplay and provide more immersive experiences, especially for applications like alternate reality and location-aware games. Alternate reality games (ARG) are essentially interactive stories that use the real world as a backdrop and that you experience through, for our purposes, some sort of mobile gadget. They encompass a wide range of game types (more than I can discuss here), but my current favorite is *Star Wars Arcade: Falcon Gunner.* Hold your camera-equipped smartphone or tablet up, start the game, and you're in the Millennium Falcon's quad laser turret, blasting TIE fighters over whatever real-world scene your camera's pointing at—the Manhattan skyline or downtown Tokyo, for example. If you've ever used your iPhone to take a photo, picture that scene and then imagine Imperial ships swooping over it, and you'll get the idea.

Location-aware games (really a subset of ARGs, but a very cool one) use your mobile device's GPS service to pinpoint your whereabouts, and then set your gaming location to your physical location—Paris, Boston, Malta, wherever you are on the globe. Take the game *Shadow Cities*, for example. *Shadow Cities* pits two teams against each other in a quest for global domination. But the world you're trying to take over isn't some sci-fi alien planet, it's Earth—and if you're in, say, London, then your job is to take London for your team and defend it against the opposing force. And you're not alone. Nearby players appear on your screen in real time; if they happen to be friendly, they may assist you. If not, you could very well find yourself under attack. However, you can set a beacon in the game at your location and draw in comrades-in-arms to help defend your position. Anyone who's aligned with you (depending on the side you chose when you first started the game), anywhere on the planet, can come to your aid. When you log off, your game-state persists, and any territory you occupy remains under your control, until an opponent comes to wrest it from you.[8]

The second facet of mobile that entices game companies is that the devices are almost always on and within reach. People carry and use them everywhere (and I do mean everywhere), and there are a lot of five-minute breaks throughout the day that many would be happy to fill with a quick bit of entertainment. And third, at nearly one mobile device per person worldwide on average (and growing),[9] the potential audience is mind-numbingly enormous. We've reached a confluence of technology, access, and reach that is unprecedented, and it's poised to bring monumental change to the video game industry.

Mobile's already revitalizing the casual games sector. Peter Hofstede of Spil Games[10] took some time out from a hectic schedule at Casual Connect to share his thoughts on this. We met at Seattle's famed Triple Door, which, along with Benaroya Hall, played co-host to the conference and also offered a few out-of-the-way tables for relatively quiet conversation. "I think it's an ideal platform for the casual games industry," he told me. "It almost seems like it's reignited this event. This event, for years, used to be about downloadable PC games for a very specific audience, and suddenly there's this new set of devices." He paused for a second. "It might be easier," he added, "for *this* crowd to get into that than for the crowd that's used to doing ten million dollar big-budget titles. It's not hard to get their heads around the fact that you need to make something small, in terms of production scale, but also in how people use it. The mindset of the core industry is quite hard to map onto mobile, and I think for the casual side it's a lot easier."

To Hofstede's point, the shift to mobile does seem to be leveling the terrain a bit between traditional big-budget console and casual games. It's put them both on an even footing and may be helping the casual industry chip away at console gaming's market share: Close to [illegible] per cent of US gamers play casual games (currently about one hundred million people), and a third of them actually use mobile games as a substitute for games on other platforms.[11] However, even though the casual sector looks poised for continued growth, Hofstede doesn't believe it'll ever *completely* supplant the console market. "I would say there's always room for big, blockbuster, hi-fidelity titles, just like big action movies are still around," he told me. But, he adds, "that's gonna be a very small group, and it seems like it's gonna be quite hard for the hardware manufacturers like Sony and Nintendo to somehow keep up with the mobile phone companies."

The primary reason behind this is simple: product lifecycle. Consoles like the Xbox, PlayStation, and Wii are replaced by their next generation offspring every three to five years. That's their lifecycle. Mobile devices have a much shorter shelf life—about a year, in most cases. So by the time the next state-of-the-art console hits the market, a typical mobile gadget's gone through three to five iterations. Given the precipitous exponential curve of technological progress, that's plenty of time to determine the course of the future. If that seems a lofty claim, bear in mind that, from the launch of the iPhone, it only took three years to build the greatest innovation in mobile entertainment history—the iPad.[12]

If there's a single gadget driving the future of mobile gaming, it's the tablet—exemplified by that most elegant of devices, the Apple iPad. Bigger than a smartphone, more portable and flexible than a laptop (and almost as powerful), the iPad represents a convergence: it's a serious piece of technology that knows how to have fun. Its combination of large, vibrant screen, fast processor, wireless connectivity, and painless mobility affords the freedom to effectively work and communicate virtually anywhere. Throw its fingertip interface into the mix, and you've got a platform that's astoundingly good for—what else?—playing games. Depending on your perspective, the advent of the iPad and its plethora of brethren at once sparked both the rebirth and demise of productivity.

On the final day of Casual Connect, I caught up with Mike Arcuri, then vice president of marketing at Zipline Games,[13] and we spoke about tablets and their influence on mobile gaming. As an appropriate preface to our conversation, Arcuri removed from various pockets and pouches a trio of touchscreen devices and laid them on the counter between us—one man's homage, at least, to the power of the tablet. Arcuri's young and energetic, and his voice conveys a natural, infectious enthusiasm. He's excited about the possibilities these devices offer, but also sees it as a double-edged sword.

"I would say there's a challenge. The gauntlet has been thrown down to game designers. How do you make a game such that it has depth to it if you want the depth? It adds more complexity, but we deal with complexity all the time. I think it can be done." As an example, he showed me one of his current favorites, a real-time strategy game called *Strikefleet Omega* that he had running on his mini-tablet (sized roughly halfway between an iPhone and iPad). Its plot is very traditional hardcore: Earth has been destroyed, and you're in command of a space fleet doing battle with the alien menace. The game looks core, even the mechanics hearken to core real-time strategy gaming: there's an economy driven by in-game currency, you mine asteroids to obtain resources needed to build your ships, and you have to make certain choices in terms of what type of ship to place and when. The difference is everything's been simplified to appeal to a noncore audience. Many casual games take a similar approach, delivering fun, quality entertainment that looks and feels hardcore yet plays casual—a near-flat learning curve, shorter time commitment, simpler gameplay.

Take *Backbreaker Football*, for example. *Backbreaker* (for iOS and Android—not the console version) is a 3D football game that has you attempting to score touchdowns while avoiding a crippling tackle. Sound and graphics are top-notch: when you're hit, there's a bone-jarring crunch and you fall to the ground with much flailing of arms and legs (tackles and touchdowns get replayed for your further enjoyment). Controlling your character is simple: Tilting the phone or using the touch-screen controls moves you either left, right, or forward as you run downfield towards the end zone. There are a couple of special moves you can execute—showboating on the way to a score is a perennial favorite. And that's it. It has all the trappings of a core game—it's 3D, graphically rich, football is a very popular genre among gamers—but the designers adjusted the skill/challenge curve down to the lowest common denominator to appeal to the widest audience. And people loved it. The game's developer, Ideaworks Game Studio, presented a keynote at the 2010 Casual Connect conference, which Arcuri attended.

"They talked about the design for this game and the analytics they put into this game, and that they refined it and refined it and refined it until the people at the office were like, 'this is idiotic. It is so easy. How could anyone possibly love it, there's no challenge here?' And that's when they had a mega-hit."

Backbreaker isn't alone in this. Casual games are, by their nature, simple—and, it seems, the simpler, the better. There's no finer illustration of this than a little distraction created by Finnish game developer Rovio Entertainment called *Angry Birds.*

Angry Birds is about as basic as a game can get. Using a slingshot, you launch an assortment of colorful birds at evil green pigs who've stolen their eggs. And that's all there is to it. Level after level, wave after wave of pigs taunt you from behind a variety of stone, ice, wood, and metal structures. Destroy the pigs, and you advance to the next level. Seems about as unlikely a formula for a hit game as you could devise. And yet, since its inception in December '09, *Angry Birds* has been downloaded almost 650 *million* times[14]—more than twice for every citizen of the United States. It's one of the most downloaded games in history, and it's massively profitable: In 2011 alone, it generated Rovio $106 million in revenue,[15] and it's spawned numerous spin-offs, licensing ventures (*Angry Birds* plushies, anyone?) and special editions—including a tie-in with everyone's favorite galaxy far, far away. And while *Angry Birds* is exceptional, there are a plethora of successful casual games on the market—and they've made the industry a lot of money. How much, you ask? Try this (but sit down first): in 2011, the casual games sector brought in an estimated $6.7 *billion* in the United States alone.[16] Worldwide, the number is staggering: as a species, we spent more than eighteen *billion* dollars[17] on silly, mindless games.

The question is, why? Why do we humans find these games so irresistible? Why can't we stop playing them?

One reason is simply that they're fun and exist on a level that doesn't ask too much of us. Mike Arcuri has his own ideas around this. "People are forcing those play sessions into whatever little time window they have," he told me. "It doesn't demand all your attention. Maybe there's something to that, in the same way that some people like TV or music because it's a little more passive, a little less demanding. Some of these casual games are just interactive enough that you're engaged and you're interested."

This is certainly true in my experience. It's why I can play *Bejeweled Blitz* or *Yahtzee Adventures* or a slew of others and still handle a phone call. I don't have to think about the game at all, I can let it proceed and give it only a fraction of my notice. I'd never try this with *Halo* or *Metroid*; those games require a level of focus that precludes any activity more cognitively intense than breathing.

Of course, some casual games are more absorbing than others. Games like *Mahjong Towers* or *MathDoku* require a bit more focus and can keep you engaged for hours. They can even bring on a relaxing, almost meditative state, which is another reason why people play them: for many, they work like pressure safety valves, providing much-needed stress relief.[18]

Casual games can also offer rewards that real life, many times, does not. When I spoke with Arcuri, I mentioned that *Tiny Zoo* was one of my son's favorite iPad games and that I enjoyed it as well. It's a very cheerful game, populated with lots of adorable animals that appreciate the care you give them, and there's almost always a new animal or reward waiting for you at the beginning of each play session. He suspected that games like *Tiny Zoo* provide regular positive feedback, frequent rewards, and a sense of control that, in the real world, many of us lack. "Real life doesn't work that way," he said. "You don't come home from work everyday feeling like, 'Man! I got promoted again today! My salary went up by two thousand bucks!' In a game, your income keeps going up and up. Because it works a little different than real life, I think one motivator is just 'man, I feel good when I do this because it's a little simpler than reality, and I can make progress and feel good about my progress.'"

This desire for achievement is one of the aspects that makes games like *Tiny Zoo* or Zynga's *FarmVille* so compelling: Their reason for being, their entire existence, is to help you build—a farm, an empire, a civilization—and to give you a sense of accomplishment. Okay, money might have something to do with it as well. After all, it's not unreasonable to think that game developers would like to make a little profit from their labors, and our need to advance provides the perfect mechanism. Once we've built something in one of these games, we become attached to it, place value on it, we want it to succeed—and while we can do this just by putting in time ("grinding," in gaming parlance), if there's a virtual item or ability we can purchase that helps speed things along, we'll gladly drop a dollar or two of real-world cash here and there. When I ran a zoo in Blue Fang's wonderful (and sadly now-defunct) *Zoo Kingdom*, I'd occasionally buy a rare animal or building. They never cost much and made the game more enjoyable. The casual sector has a word for this: microtransactions. And though each individual purchase typically doesn't amount to much ($1.50 is around average), in aggregate they account for about 60 percent of a casual game's revenue.[19] Microtransaction purchases range from custom game backgrounds and avatar outfits, to special services and game power-ups, to limited-time items like special animals, tools, or holiday-themed articles—a virtually limitless variety, none of which actually exist.

Not that it matters. Virtual products do a brisk business: From November 2006 to the following June, players spent $6.7 million on virtual gems to upgrade games through game portal Pogo.com. In 2006, game developer King.com reported microtransactions totaling $27 million. *Habbo Hotel* makes 90 percent of its revenue ($77 million in 2006) from microtransactions and in 2007, the game sold more furniture than Ikea. And in one month in 2007, North American gamers bought $1.6 million worth of virtual products in *MapleStory*.[20] Though the items are virtual, the money's as real as the reasons these games capture us: enjoyment, stress relief, reward, a sense of accomplishment.

Peter Hofstede believes there's a deeper reason—something that gets people hooked on games in general:

> We have certain needs. Evolutionarily, we have certain things we like to do. Some people, they enjoy collecting, for example. A certain lot of these casual games are these match three games, and there's a theory that it's the idea of cleaning the board that's appealing on a very deep, instinctive level. I think that what casual gaming did is it brought these mechanics to a new audience, an audience that previously wasn't playing games. And of course they like these games. It's human nature to like this kind of thing.

Human nature. Exactly. Gaming is fundamental to our experience as human beings, to our biology. Like the need for food, shelter, clothing, and love, the need to play is coded in our genes. Who's a gamer? We all are. Whether our game is played with dice, cards, tiles, pencil and paper, on a board, or on the latest touchscreen tablet, we all have one, we're all linked by the common thread of play.

Why are casual games so compelling? Because we are, above all else, human. And because play, in any form, is essential to our humanity.

6

DRESSED FOR THE SYMPHONY

Video Games Take Center Stage

April Fool's Day, 2010. New Orleans. The Mahalia Jackson Theater. Out of a packed house, a young boy emerges. He takes the stage, straps on his guitar—cool, cocksure—and waits for his cue. The conductor lowers his baton and the opening bass guitar/talkbox duet of Aerosmith's "Sweet Emotion" fills the theater. The boy takes a deep breath, releases, and then tears into the song, backed by a rock band, a thirty-four–person chorus, and the Louisiana Philharmonic—and in front of more than two thousand screaming fans. His concentration is total. He stands rooted in place, his fingers betraying only the barest hint of effort as they grind out fretboard runs with precision, grace and fire. His performance is flawless.

And it's all a game. Guitar Hero, *to be exact. And the guitar? It's a video game controller, with buttons on the neck instead of strings. The boy's playing the game live, as part of the New Orleans stop on Video Games Live's world tour. He won the honor by logging high score during the symphonic music concert's preshow* Guitar Hero *contest. Now—in front of a capacity crowd, on stage with professional musicians who've been playing longer than he's been alive—the kid has to score at least two hundred thousand points on the game's hard skill level.*

"Expert," he calls out without hesitation, catching the co-host by surprise.

"What? You want it on Expert?" he asks, the tone of his voice adding an unuttered "You sure about this, kid?"

Yeah, he's sure. Expert it is.

The audience eats it up. He had them at his first step to the aisle, and now they're hanging on every note. The kid hits a fifty-note streak, then one hundred. The crowd cheers. His score passes 150,000. The co-host screams "150!" and the audience responds in kind. Then 160, 170 . . . the theater vibrates with collective energy. It's getting near the end, but he's cool. 190. The audience is going wild, nearly on their feet. 195. They're barely holding it together. 196 . . . 197 . . . 198 . . . 199 . . . And then, over Joe Perry's climactic guitar solo, the boy clears two hundred thousand—and the audience explodes. Parents, kids, old, young, all leap to their feet as one, cheering like teenagers at a stadium rock concert. And the kid? He keeps playing through to the final brass crash of the cymbal. When it's over, he doesn't even crack a smile, just gives the conductor a nonchalant mid-five, hands the guitar back, and exits the stage.

Cool.

• • •

Martin O'Donnell. Gerard Marino. Christopher Tin. Jason Hayes. These guys may not be household names, but in certain circles they're as famous as John Williams, James Horner, or Hans Zimmer—and for similar reasons. Over the past three decades, Williams, Horner, and Zimmer scored some of Hollywood's most iconic films—films that influenced generations of moviegoers—and changed the role of movie music forever. No longer simply background to a film, the soundtrack became an integral part of the story—try to imagine the TIE fighter attack on the Millennium Falcon without the driving pulse of the orchestra, Ellen Ripley and the surviving Colonial Marines' flight from LV-426 just ahead of the nuclear blast without Horner's pulse-raising, staccatoed crescendo, or Indiana Jones' narrow escapes from danger without his signature trumpet fanfare, and you get the picture. Perhaps more significant, though, they changed the *perception* of movie music from incidental score to serious art form, something you could listen to *apart* from the movie. True, there were others before them—Henry Mancini wrote some memorable themes (who hasn't heard *Peter Gunn* or *The Pink Panther*?) and was the first film composer to score a hit song for a soundtrack—but Williams, in particular, transformed the soundtrack into music that could not only complement a film, but stand on its own merits.

O'Donnell, Marino, Tin, and Hayes have done the same for a new generation of listeners who've grown up not with *Star Wars* and *Aliens* but with *Halo* and *World of Warcraft.* Unlike previous generations, their primary choice of entertainment isn't movies, it's video games—with soundtracks as familiar and beloved by gamers today as the great Hollywood composers' work is by those of us who were kids in the '70s and '80s.

If you've been out of gaming for a while, it may seem like a stretch to compare a video game soundtrack to a film score by an Academy-award–winning composer—or to think of video games as even *having* soundtracks—but then you're probably used to the early days when most games could produce a maximum of up to eight simultaneous sounds (electronic blips and bleeps only) or, at best, a synthesized approximation of an orchestra. While this was a vast improvement over *Pong*'s single tone, it's a far cry from the rich and colored soundscapes game composers can now create.

Perhaps the best example of this is Martin O'Donnell's work on Bungie's *Halo* games—arguably the most renowned and successful video game series ever. They're regularly ranked among the best first-person shooters and have spawned a variety of related products, including books, graphic novels, action figures, building toys—and, yes, CDs. And deservedly so: The *Halo* soundtracks are astonishing works—and not just because they're written to accompany video games. Each score reflects the characters, events, moods, and environments of the game, and includes main and subsidiary themes that recur throughout the series. The music has surprising sweep and drama, passages of beauty, emotion, power, suspense, and even quiet reflection. The soundtracks all stand as discrete works, yet—as with the best multi-film Hollywood epics—when taken together, they create a coherent suite far greater than the sum of its parts.

As a kid, I spent hours listening to Hollywood's Holy Trinity: *Star Wars*, *The Empire Strikes Back*, and *Return of the Jedi*. I had favorite pieces, favorite themes (to my mind, *The Imperial March* stands as the greatest movie villain theme ever written). I knew every cue and could tell you with 100 percent accuracy *exactly* what was happening on-screen during a particular musical passage (and if there was dialog, I could give you that, too). And I wasn't alone: virtually all of my friends—and, in fact, most kids in my age group—could as well.

Now, I grew up listening to a lot of music—Bach, Beethoven, rock and roll, metal, pop, opera, jazz, the blues, Broadway—and developed an ear for appreciating the sublime, regardless of genre (I mention this merely to make the point that, though I enjoyed a wide range of music, I did not do so indiscriminately). I loved classical music, but during my tween/teen years in the early and mid-'80s, classical carried with it a certain geek/nerd stigma—and I didn't need any help in that department. But John Williams was different. At a time in my life when listening to classical music wasn't cool, John Williams was validating. I could put on one of his soundtracks, and if another kid made a face or a comment, all I had to say was, "but it's *The Empire Strikes Back*," and the kid would relent—because the movies were *cool*. Everybody loved them, and for kids of that generation, they defined us.

Video games today do the same thing. A popular game or game series is as culturally significant to gamers now as movies were thirty or forty years ago. As a result, many tie-ins become cool by association—including the music. You gamers know what I'm talking about. The rest of you probably think I'm out of my mind. While you're most likely willing to grant me the validity of movie music as an art form, I can hear your game-music-as-music objections loud and clear—something along the lines of, "but film music is written by *real* composers, performed by *real* musicians—lofty and much-lauded orchestras, even. It has emotional depth and range. And it's regularly performed *live*. For an audience. With *real* musicians."

Okay, yes. All true. But the same can be said for video game music. Verbatim.

Regarding musical pedigree, some of the game industry's finest composers studied at institutions like Stanford University, Oxford, London's Royal College of Music, Wheaton College Conservatory, Mannes School of Music, and the Universities of Southern California, Michigan, and North Texas. When composing for a game, they speak of creating an overall sound evocative of the game's mood, environment, and timeframe, building themes that capture the protagonist's motivations and struggles and reflect the game's story arc and emotional range sonically.[1] And they regularly mix live musicians in with synthesized elements—sometimes eschewing synths entirely, opting instead for the quality and depth of human performance.

Consider, once again, Martin O'Donnell's music for the *Halo* trilogy, which he composed with long-time collaborator Michael Salvatori. The first game in the series, *Halo: Combat Evolved*, finds them combining keyboards, synthesizers, and samples with live performances from members of the Chicago Symphony and Chicago Lyric Opera Orchestra. For *Halo 2*, they took it a step further, enhancing sampled electronics with Seattle's Northwest Sinfonia orchestra and spicing it up with a judicious amount of prog-rock guitar hero Steve Vai's incendiary fretwork. *Halo 3*—the main trilogy's finale—saw O'Donnell and Salvatori returning to and reworking selected themes from the previous installments, giving them new life with orchestral treatments, as well as composing new themes to convey the game's sense of scale and resolution. The biggest difference, though, was that this time around, they recorded the *entire* score with the Northwest Sinfonia and a full twenty-four–voice chorus. The result? In a word: music. And not just music that perfectly complements the game (although it does; some reviews went so far as to say that *Halo 3* wouldn't be the same game without it); music that you'd actually listen to even if you'd never played any of the *Halo* series. I've only played through *Combat Evolved* myself, and I've been spinning the soundtracks for the last several days—and not just out of professional curiosity. The music is captivating, compelling, full of bombast and subtlety and surprising emotion. And it's for a video game trilogy. I had to keep reminding myself of that. If I'd been listening blind, with no knowledge of the source, I'd have thought it was the score to a Hollywood blockbuster. My wife felt the same way. Clearly, this is the real deal.

Just to be sure, though, I tried it out on our friends Bethany and Joe—casual gamers at most, with only a peripheral knowledge of *Halo*, if any at all. All I asked was that they really listen for a few minutes and then give me their impressions. They both picked up on the themes, the sense of the music telling a story, building towards something. Bethany was especially drawn-in—eyes closed, intent. In terms of emotional range, for her the music evoked *Les Miserables*. Both could see themselves playing it at home, and both were completely taken aback when I revealed the source.

Again, it's not shocking that my peers were stunned by the state of music in games today; nothing in our experience prepared us for its variety and breadth: Gregorian chants, world music, classical orchestras, full choirs, rock, pop, blues, jazz, hip-hop, punk—it's all out there. In some ways, though, we should have seen it coming. Video games have always pushed the envelope of the possible, driving advances in technology—better graphics, more power, speed, and capacity—which, in turn, led to further artistic and technological leaps in games. Whether the games are any *better* is arguable; that they look, sound, feel, and react better is not. It was only a matter of time before composers, musicians, and game developers realized what they could do and ran with it. Freed from the shackles of eight-tone electronic instrumentation and woefully inadequate machine memory, they could stretch out, flex musical muscle, and give their imaginations virtually unlimited rein.

I asked Andy Brick about this on a June morning at his home a few years back. Brick's been a game music composer for more than fifteen years and a film composer for longer. Among the games he's scored are *Kingdom 2: Shadoan* (sequel to *Thayer's Quest*, direct descendant of the groundbreaking *Dragon's Lair*), *Sim City 4*, *Stranglehold*, *Merregnon 2*, and *The Sims 2*. At the time, he was also the conductor and music director of the Symphonic Game Music Concerts in Leipzig, Germany (more on that later). When he started writing for games, samplers were all the rage: they allowed you to record real instruments into an audio library, which you could trigger with a digital keyboard hooked to your computer and then use in scoring. If you wanted a violin, you'd simply load the violin sound sample into your keyboard and then access it through the keys—just like playing a piano. Samplers made the music sound more real but still had their limitations. No matter how good the samples were, or how skilled the composer, to Brick's ear, the music still lacked a human element. "There's so much that a live person does that the samples and synthesizers can't," he told me. Triggering a sample gets you a single note, "a static moment in time," but it doesn't give you the human *performance* of the note, or the emotional experience and reaction of the orchestra to the way notes are being played, to the music. From the outset, Brick took a different approach.

> And my thing was that when I started doing games, I always insisted on having at least one live instrument. And before a few years, I started to have small ensembles for games. I did this game called *Kid Pix Deluxe*, which was a kid's game, and wound up . . . at that time, the technology was such that we could do overdubs, where we could record a single violin, and then record him again, and then record him again, and build these huge orchestral sections. That game, I think I had a budget for eight players, and I just kept overdubbing and overdubbing, and built a seventy-piece orchestra that way.

By using a bit of audio trickery—combining multiple performances of a small group of musicians with sampled sounds from his library—Brick captured some of the nuances of human musical expression and created the illusion of a live orchestra. It's a technique used by many composers to balance the desire for high-quality music against game budgets too limited for a full orchestra: when money is the obstacle, technology provides the path around. Regular and rapid leaps in the quality and capability of electronic tools afforded composers the freedom to experiment and gave them access to sounds and instruments that might have otherwise been beyond their reach—and resulted in some wonderful and innovative music.

But then a curious thing happened. As game music started sounding better, developers and gamers began demanding better still. When composers delivered, the audience again asked for more. Ironically, the more audio technology made it easier for composers to replicate a live orchestra, the more people demanded . . . a live orchestra. From Brick's perspective, it was a natural progression, an evolution. "As the computers and synthesizers and samplers are able to create more realistic sounds," he told me, "the audience's ear has become more sophisticated and more demanding. Well now, the audiences have been saying 'give me more' to the point where they now need the real orchestra, the fake just doesn't do it, and you hear it immediately, the difference between the two."[2]

For composers skilled at scoring for a live orchestra, this shift is a godsend.[3] It's gotten their music into the hands—and ears—of a new, very large, and extremely enthusiastic audience, provided fantastic creative opportunities in an energetic, growing, and as yet unsaturated industry, and given birth to new musical careers while rejuvenating others.

The push for live orchestral music also had an unexpected consequence: it sparked a convergence between the formerly disparate worlds of symphony and video game. This convergence culminated in 2003 as the first Symphonic Game Music Concert, held at Gewandhaus in Leipzig, Germany, performed by the Czech National Symphony Orchestra, and conducted by Andy Brick.

The first program in the world to feature video game music—the Orchestral Game Music Concert—was actually launched in 1991 by Japanese game composer Koichi Sugiyama. This very successful five-concert series ran from 1991 to 1996 and was performed by a variety of orchestras, but only toured in Japan.[4] A concert of video game music had never been tried in the West, and no one knew what to expect. As the theater gradually filled to capacity, though, there were the first hints of success. The makeup of the audience was another indication: gamers showed up, to no one's surprise, but so did their parents, and even grandparents—teens to seniors all coming together for a night at the symphony. It was likely the widest age range for an audience of symphonic music ever. Their enthusiastic response, capped by a ten-minute standing ovation and four curtain calls for the orchestra, confirmed the show's success: video games were ready for the concert hall. Brick went on to conduct four more Symphonic Game Music Concerts—all to sold-out crowds. Following the 2007 concert, Brick joined PLAY! A Video Game Symphony, first as Associate Conductor and now as Principal Conductor and Music Director, and continues to perform game music to capacity crowds and standing ovations. Against the odds, orchestral music had found appreciation in the most unlikely of places: gamers. And the parents of those gamers were beginning to understand, for the first time, the power, depth, and beauty of video game music.

This burgeoning mutual appreciation wasn't lost on long-time game music composer Tommy Tallarico. An industry veteran of close to a quarter century, with more than 275 game soundtracks to his credit, game music's leap to the symphony was something he'd been anticipating and suspected might be successful. Audiences were beginning to value game music with a human touch, and Tallarico was in a position to deliver. He conceived of a live music experience that would combine the excitement of a rock concert, the emotional range of a symphony, and the "cutting-edge visuals, technology, interactivity, and fun"[5] of a video game. Actually planning and producing a live game music concert, though, was another level altogether. Tallarico needed a partner. Trouble was, nearly everyone thought Tallarico was, in a word, insane.

Nearly.

Enter Jack Wall—composer, conductor, and, according to everyone but Tallarico, fellow insane person. Wall loved the idea, and together, he and Tallarico spent three years planning the event. Video Games Live debuted at the Hollywood Bowl on July 6, 2005, with Jack Wall at the podium, the Los Angeles Philharmonic on the stage, Tommy Tallarico on electric guitar, and eleven thousand game fans, friends, and family in attendance.

It was a rousing success. The tour played two more sold-out shows that year, both with the Northwest Sinfonia. In 2006, it expanded to eleven shows worldwide. In 2007, the tour nearly tripled to thirty shows, climbed to forty-seven in 2008, and peaked at seventy in 2009. Since 2009, Video Games Live has been running, on average, around thirty shows a year (though there are fifty planned for 2013), many to full houses.[6]

Part of the show's success lies in its production. While PLAY! and Video Games Live both feature top-notch orchestral arrangements and musicianship, PLAY! feels more like a traditional evening at the theater. It's predominately an auditory experience: Though there are video screens, they're just as likely to show the orchestra as they are to display game footage. And there's a notable lack of electric guitar. Video Games Live is PLAY!'s less refined, edgier sibling—the Jimi Hendrix to PLAY!'s George Benson. A Video Games Live concert is a multimedia spectacle. It's triple-A symphonic music accompanied by synchronized lighting and video, lasers, a rock band, appearances from popular video game characters, surprise guests, on-stage audience participation, and live action. It is grand theater.

But it's more than that. Video Games Live is an *event.* Before each performance, attendees can meet different composers and game designers and take part in interactive demos, a costume contest, and various video game competitions—including a *Space Invaders* contest that pits parents against their kids and a *Guitar Hero* match that gives the winner a chance to play the game on stage during the concert, backed by the entire orchestra. Some stops on the tour also feature an exhibit that guides visitors through the history and evolution of video games.

Taken together, it imparts a distinctly festival air to the proceedings. Add this to the multimedia extravaganza and the dramatic symphonic music, and it creates an experience that resonates with audiences of all ages.

Shelly Fuerte, artistic administrator for the Pittsburgh Symphony Orchestra (the host orchestra for Video Games Live's Pittsburgh stop), sees it as more than simply bringing different generations together, though. "I don't think the older generation knows that kids are really interested in serious music," she said. "Gamers know the quality of the music and they go crazy when they hear it done live, but it's the non-gamers who are really blown away by the whole presentation."[7]

It's a phenomenon that Tallarico's experienced often: Some of the most enthusiastic letters he gets come from nongamers who found themselves affected by the music, much to their surprise. Parents, too, send grateful messages, thanking Tallarico and Wall for producing the concert series. For them, the benefits are two-fold: they gain an understanding of a pastime that's important to their kids, and it gets their kids excited about going to the symphony and hearing a style of music that's important to the parents.

Making a live orchestra appeal to kids—in short, making it cool—has historically been a Sisyphean task, to the dismay of parents and music teachers everywhere. Video games are changing that. After one of the Symphonic Game Music performances, Brick received an email from the mother of a teenage boy who'd been at the concert. "She said, 'I just want to thank you. My son came back from the concert and immediately asked me if he could go see another violin concert.' This mother was expressing gratitude because there probably was very little opportunity for her to convince her son that this is a viable form of music." Now, however, there's an orchestra playing music from a genre he knows and loves, and "voilá! Now . . ." and Brick dropped his voice to a hush, ". . . he's kind of into classical music."

This is not an uncommon story. For many kids, these concerts act as a kind of gateway drug into the larger world of the symphony; after hearing an orchestra perform their favorite game music, some are moved to discover what else the orchestra has to offer. From Mario and Zelda to Wagner and Shostakovich isn't that far a step, after all, and it's not a surprise that this newfound appreciation for symphonic music would carry over to more traditional material. Says Tallarico, “We’re also helping to usher in a whole new generation of young people to come out and appreciate the symphony and appreciate the arts.”[8]

And not a moment too soon, as it turns out. Orchestras, you see, are in real trouble. When asked about this, Brick doesn’t mince words.

> I mean thank God for the sake of the orchestras that there’s something out there that’s bringing these younger audiences back to the concert halls, because orchestras are dying, they still are. When you go to a rock concert, you go ’cause you like to sing along. You don’t have as much of an urge to go to a rock concert if you don’t know the songs that they’re gonna play. Well it’s the same way in a symphonic concert. If you go and you tell them you’re doing the theme from *Zelda*, you know, everybody is “oh, I know that! I played that game for thirty-five thousand hours, and I want to go check that out, that’s gonna be great.” So they buy a ticket to the concert, and all of a sudden, the orchestra is engaged and they’re playing for, not only a live audience again, but they’re playing for an audience of twenty-five–year–olds. . . . A lot of smaller regional orchestras are starting to fill their concert halls with younger kids now because they’re playing this repertoire. It’s really great.

Some kids take this new love even further. Not content simply to listen, they want to participate in the music’s creation. Brick’s talked to kids from grade school to high school, and when he plays them music from the games they know, they respond. “It makes them want to play as well,” he told me. “They like playing the game, and they want to be a part of the art. And the art in this case is the orchestra, and so they want to be a part of the orchestra.” It’s exactly the same as when I was growing up and heard *Star Wars* or *Raiders of the Lost Ark* for the first time: I wanted to *play* those themes. Video games today are doing what movies did thirty years ago—they’re getting kids excited about taking up an instrument and playing music. And that’s helping to save orchestras from the inside, filling seats with young, enthusiastic musicians eager to share their love of music and inspire the next generation.

From its modest beginnings as the single-tone blips of *Pong* and the incessant, oppressive drone of *Space Invaders* and *Asteroids* to the sampled orchestras and live symphonic performances of today’s best games, video game music has done the extraordinary: it's transcended its original purpose and become what many forms of expression only aspire to be—true art.

7

FROM THE FLAT SCREEN TO THE BIG SCREEN

Video Games Invade Hollywood

It's opening night. Fans have been anticipating this moment for months, and now the wait is finally over. The line extends down the street and around the corner, filled with those eager to be among the first in the world to experience this event—the finale, the conclusion, the capstone to a trilogy they've known, lived with, and loved for years. They've been standing here for hours, swapping stories about their first exposure to the series, sharing favorite moments, debating merits and enthusiastically discussing the finer points of each installment. The most hardcore among them passionately dissect the individual chapters—character development, visual effects, music and sound, writing, acting, plot and pacing, storyline, immersion. The mood is electric, the air around the gathered throng seeming to vibrate at a higher level, molecules moving to an alien quantum tempo.

And then the doors open, and the masses move towards the entrance as one, compelled to motion by some invisible, collective force, threatening to overwhelm the staff by sheer numbers and the insistent press of ardent humanity. From the head of the line comes the telltale ring of the register, and then one lucky fan holds up his prize, face aglow with pride and expectation: the first-purchased copy of Halo 3. *It's not the latest Hollywood blockbuster that's drawn this crowd, but the official release of the most highly anticipated video game in history and cur-*

tain-close to the Halo *trilogy, the most beloved and heralded video game series on Earth.*

• • •

When I was a kid, a group of friends and I waited in line for hours to catch the premiere of *The Empire Strikes Back*. It had been three years since Darth Vader escaped to an uncertain fate after Luke Skywalker's shot heard 'round the galaxy, and from that point on, until we started seeing previews confirming the Lord of the Sith's survival, the question on all of our minds was, *what happened?* None of us was old enough to remember television's golden age of weekly cliff-hangers (Will the hero survive? Will the villain rise again? Tune in next week to find out!)—that was our parents' generation—and so the idea of a sequel was, for us, uncharted territory. Yes, the James Bond and Pink Panther movies had some recurring characters, but they weren't proper sequels: Subsequent films in those franchises rarely (if ever) carried on the storyline from previous films—and you could watch them out of sequence or skip movies you didn't like without really missing anything.

This was different. First of all, *Star Wars* had a cultural impact far greater than either 007 or Inspector Clouseau could have dreamed—arguably greater than any film since movies first came to life in the late 1880s. When *Star Wars* hit the silver screen in 1977, it represented a clear demarcation, dividing the history of modern cinema into everything that came before and everything that followed. And it marked a profound shift in both the industry and the audience—in a very real sense revealing entire new worlds of possibility, constrained only by the limits of imagination. When George Lucas released *Empire* three years later, it was the first movie my friends and I had ever seen that picked up its predecessor's thread and drew it forward, spinning it out beyond anything we had imagined. Old conflicts were rekindled and deepened, long-lost secrets revealed. The characters we loved had been set on new paths, forced to confront new and greater challenges. To us, these were matters of weight and consequence. We *cared* about these characters, we celebrated their triumphs and agonized over their defeats. Perhaps most important, though, we felt a certain proprietorship of *Star Wars*: The movies were created for *our* generation, they belonged to us. We were their target demographic, and they spoke to us in a way that resonated. It didn't matter if anyone else understood them. In fact, it was better if they didn't; it made them uniquely ours. And it wasn't just the movies. *Star Wars* was a unifying experience, something that cemented our generation in a particular place and time, provided a common point of reference—and something that those of us who grew up in the *Star Wars* age still talk about, nearly four decades later. Every generation has a similar event. For my parents, it was The Beatles' appearance on *The Ed Sullivan Show*. For me and my peers, it's *Star Wars*. For the generation that follows, it may very well be *Halo*.

For the uninitiated, *Halo* is a video game series comprised (as of January 2013) of five main games plus two spin-offs. The games are regularly hailed as among the best first-person shooters ever produced, and—in a nod to their cultural influence—have been dubbed "*Star Wars* for the thumbstick generation."[1] It's a fitting description: *Halo* fans today—known collectively as the "Halo Nation"—are as passionate and eccentric as their Force-obsessed elders. They've played all the games (repeatedly), own the soundtracks, read the books, and bought the T-shirts and figurines. They discuss the intricate details of each game with a fervor usually reserved for religion or politics, and pride themselves on their knowledge of arcane facts and plot points. Much of this information comes, not from the games, but from a variety of media adaptations, including more than a dozen novels (many of them best-sellers), half a dozen graphic novels and comic books, a live-action web series (*Halo 4: Forward Unto Dawn*), a collection of anime short films, and a spate of guides and art books.[2] Descendants of tomes like *The Art of Star Wars* and *The Empire Strikes Back Notebook*, these books contain the entire history of the *Halo* universe: concept art, creatures, vehicles and technology, characters, architecture and environments, and models and renderings. There's also a fair amount of commentary and description, giving the reader insight into the development of each game's visual style and what the artists were trying to convey in different scenes, levels, and locations. And the guidebooks aren't simply strategy manuals for beating the games (though there are those as well), they're encyclopedias that lay out *Halo*'s backstory, delve into character history, provide a comprehensive timeline, and describe every item, vehicle, weapon, alien, artifact, planet, and human character that makes an appearance anywhere in the series. These are serious works, and reading them is very much like peering behind the curtain of Lucas' epic space opera.

In its success, the *Halo* trilogy invites yet another comparison to *Star Wars*. When 20th Century Fox released the first film, it was a crap-shoot. No one knew if *Star Wars* would be a triumph or a complete flop—and no one could have predicted the level of success it would ultimately reach. So, too, with *Halo: Combat Evolved*. With a budget of thirty million dollars, at the time it was unveiled (November 15, 2001) it was one of the most expensive video games ever produced. Once released, it made history again, selling one million units faster than any next-generation console game before,[3] racking up accolades from sources ranging from popular video game news magazine *Game Informer* to the Academy of Interactive Arts & Sciences, and garnering rave reviews from entertainment media and fans worldwide.[4] In the annals of gaming, it was unprecedented—and it fundamentally changed the way video games became promoted and released. These days, game launches are often preceded by campaigns that take advantage of many Hollywood marketing staples—movie trailers, TV spots, glossy, full-page magazine ads and special websites[5]—and have budgets in the tens of millions of dollars (some reach one hundred million dollars or more). The releases themselves are high-profile affairs, attended by news media and local celebrities. Many are catered by local restaurants and offer complimentary food and beverages to the expectant masses. Microsoft's *Halo 3* debut was accompanied by events around the world, featuring product giveaways, gaming tournaments, and appearances from Bungie[6] team members (the game studio behind *Halo*) and professional athletes. Miami's Circuit City—one of four US locations selected by Microsoft to host the biggest launch-day parties—had then-current Miami Dolphins stars Chris Chambers and Joey Porter on hand to sign autographs and challenge fans to *Halo 3* matches. Across the pond, music superstar and self-professed hardcore gamer Pharrell Williams emceed the *Halo 3* premiere at London's BFI IMAX cinema. Similar events took place across Europe in cities as far-flung as Paris, Madrid, Amsterdam, and Milan. All told, more than ten thousand US and one thousand European retailers held midnight openings for *Halo 3*. At the time, it was the most successful release in history.

And not just for a video game. For *any* entertainment property. Ever.

How is this possible, you ask? After all, in the entertainment world, Hollywood is king, right? Movie studios bring in the *serious* bucks. Ask the average person which makes more money, movies or video games, and he or she'd likely reply with some variation of "movies, of course." They'd give the answer quickly and with conviction. And they'd be wrong.

If you're familiar with the video game industry, this probably comes as no surprise. But for the rest of us, it's almost as big a shock as learning that—spoiler alert here—Darth Vader is Luke's father. Consider this: *Halo 3* made $170 million dollars in the United States *alone* in its first twenty-four hours of public existence. Most Hollywood films look to break the one-hundred-million-dollar mark on their opening weekends,[7] and *Halo 3* surpassed it in a single day. The most successful opening day in Hollywood history (as of the end of 2012), *Harry Potter and the Deathly Hallows Part 2*, made ninety-one million dollars, just over half *Halo 3*'s haul.[8] And this wasn't the first time: *Halo 3* claimed the first-day sales title from its predecessor, *Halo 2*, which brought in $125 million—and it was unseated three years later by its descendant, *Halo: Reach*. And before you accuse me of being biased towards *Halo* (I am, but that's another story), there have since been more successful games. Let's take another look at 2011's best-selling film, *The Deathly Hallows Part 2*. Its domestic gross for the year was just over $381 million; by contrast, Bethesda Softworks' open-world role-playing game[9] *Skyrim* (also released in 2011) made $650 million in its first month.[10]

By far, though, the heavyweight title goes to the latest installment of Activision's rampantly popular *Call of Duty* franchise: *Black Ops II*. Before I give it away, allow me to keep you in suspense a bit while I set some context. Along with topping the chart for opening-day sales, *The Deathly Hallows Part 2* also holds the record for best worldwide opening weekend at $483 million.[11] *The Avengers* rang in the biggest domestic opening weekend ever—$207 million—and passed the $600 million global box office benchmark in twelve days.[12] Congratulations, both. Achievements to be celebrated, without doubt. Nonetheless, these numbers pale in comparison to the performance of *Black Ops II*. On its opening day (November 12, 2012), this video game posted a record-demolishing five hundred million dollars in global sales.[13] Fifteen days later, it crossed the one billion dollar mark[14]—only three days after *The Avengers* hit just north of $650 million worldwide.

Okay, so a few games are eclipsing Hollywood blockbusters. What about the industry as a whole? On the grand scale, are video games really dominant, or does the time-honored silver screen take the gold?

While exact, long-term comparisons between individual movies and games can be hard to make and are somewhat tenuous (movies have a longer productive life during which to bring in more money, but games carry a heavier price tag and can be more profitable at the outset), when you look at overall industry performance, the story sharpens into focus: Industry-wide video game revenue for the last several years has bested the Hollywood box office by more than double, at home and abroad. Domestically, 2011 movie ticket sales totaled $10.2 billion,[15] against video gaming's nearly twenty-five–billion–dollar take.[16] Globally, the picture looks the same: while moviegoers spent almost thirty-three billion dollars at the box office,[17] gamers dropped sixty-eight billion dollars[18] on their entertainment of choice. The fact is, annual video game industry revenue surpassed Hollywood years ago, and Hollywood's never caught up.

Video games aren't just encroaching on Hollywood's market share, either. They're subsuming the dialect of film. Long the domain of the big screen, concepts like cinematography, writing, sound design, acting, plot, and scoring are being applied to the game screen as well. In 2009, *The New York Times* columnist Seth Schiesel praised Sony's *Uncharted 2* for its use of camera angles, pitch-perfect writing and voice acting, and sound design (both the game's score and sound effects).[19] "For decades," he said, "video game makers have suffered in the inevitable comparison with their Hollywood counterparts . . . movies have had the interactive world just plain beat when it comes to sophistication of scope, characterization and visual storytelling."[20]

Not anymore. With *Uncharted 2*, Schiesel said, game designer Naughty Dog "absorbed the vernacular of film and then built upon it productively, not slavishly, to create something wondrous."[21] Upon completing the game,[22] Schiesel was left with the feeling that it might be a long time before he was treated to an action-adventure movie as compelling.

Thanks to leaps in technology over the last five to ten years, game designers now have the technical resources to bring their wildest visions vibrantly to life, and many are producing video games of unprecedented sophistication, power, and beauty. Like Schiesel's experience with *Un-*

charted 2, the best games today are beginning to rival Hollywood blockbusters in terms of audiovisual fidelity, depth, story immersion, and sheer entertainment. We're entering an age of movie/video game convergence that we've slowly approached for decades, but has, until now, been tantalizingly out of reach.

• • •

Like many relationships, early liaisons between movies and games were rocky and tumultuous. One of the first attempts to bring these worlds together was 1982's *E.T. the Extra-Terrestrial*. Based on the year's most popular movie and shepherded by two entertainment visionaries—Steven Spielberg and Atari, Inc.—the game should have been a success.

It wasn't. Quite the contrary, actually: it was an unmitigated disaster.[23] *E.T.* is widely cited as one of the worst video games ever made[24]—a mark of distinction in its own right—and is among the most spectacular commercial fiascos in gaming history. The game cost $125 million to make and another twenty to twenty-five million dollars just to acquire the rights. In a move of extreme and ill-advised optimism, Atari produced some five million *E.T.* cartridges—only 40 percent of which sold, resulting in a one-hundred-million-dollar loss. The rest ended up in a landfill in New Mexico.[25] Fallout from the game was so bad that it's regarded as a contributing factor not only in Atari's 1984 bankruptcy but in the entire video game industry's massive crash in 1983.[26]

The industry has obviously recovered from that low, and over the last two decades, many movie franchises have been attended by successful game releases. James Bond, *Star Wars*, *Spiderman*, and *Harry Potter* have all had critically acclaimed and well-received games developed that either follow the movies or extend their plots with original stories and new characters. Nowadays, video games have matured to the point where, for top-grossing movies at least, a game release is the rule rather than the exception.

Translating video games into movies also produced some incredible flops—the most notable of which was *Super Mario Bros.* Released in 1993, it quickly went on to underwhelm critics and moviegoers everywhere. Gene Siskel and Roger Ebert—wielders of the almighty thumbs—panned the movie on *Siskel & Ebert at the Movies* and added it to their list of the worst films of the year. In his review, *The Los Angeles Times*' Michael Wilmington took the script to task. "If you judged it on the writing," he said, "your final score would be feeble."[27] And Bob Hoskins, who played the titular Mario, called the film "the worst thing I ever did."[28] Ultimately, the audience had the final say: *Super Mario Bros.* made only twenty million dollars at the box office, recouping less than half of its forty-two–million–dollar production budget.[29] There have since been hits, of course—*Tomb Raider*'s Lara Croft (played by Angelina Jolie) has adapted well to the big screen, and the five *Resident Evil* movies, starring Milla Jovovich, have all scored favorably with audiences[30]—but the path to success is littered with the corpses of their less fortunate brethren, and though some have become popular and financial hits, critical acclaim is largely elusive.

If Hollywood's struggled to create a standout game movie that's achieved recognition or legitimacy, video games have still managed to infiltrate the film world and assert their cultural significance—*Tron* being the earliest example. Released in 1982—the tail end of video games' golden age—*Tron* was one of the first movies to use computer graphics extensively and the first to incorporate the lexicon of games and gaming into its core story. It didn't arise out of a video game, but they figure heavily in the movie's plot: Kevin Flynn (the film's protagonist, played by Jeff Bridges) is a software engineer who designs video games in his spare time and finds himself trapped inside one of his own creations. In order to escape, he must fight his way out by competing against programs on the Grid—a virtual arena that features many video game–style contests and where the losers are "derezzed." *Tron* met with generally positive reviews, made almost twice its production budget at the box office, spawned a slew of video games, and influenced several key figures in entertainment—one of whom was Pixar cofounder John Lasseter, who credits *Tron* with inspiring the first film in a record-setting animated trilogy and the first feature-length movie done entirely through computer animation. "Without *Tron*," he said, "there would be no *Toy Story*."[31]

The 1980s saw several other films that either featured video games as a plot point or referenced them as a cultural phenomenon—*WarGames* (1983) and *The Last Starfighter* (1984) among them—and they've appeared in countless movies since. Most recently, video games took center stage in 2012's *Wreck-It Ralph*, a charming movie that explores the lives of video game characters after hours, when they're not working in their respective games. There are abundant references to classic '80s arcade games, the story is engaging and well-developed, and the movie proceeds with effervescence, joyfulness, and a surprising amount of heart. Still, *Wreck-It Ralph* shows its influences clearly; gaming is at the fore throughout.

Go back to the summer of 2010, though, and you'll find a movie whose gaming imprint is more subtle. It features an A-list ensemble cast and a writer/director who's been hailed as one of his generation's most accomplished filmmakers. It's won four Academy Awards and, at over eight hundred million dollars worldwide, is within the top fifty highest-grossing films of all time. And it doesn't mention video gaming once. The movie is Christopher Nolan's mind-bending *Inception*, and though it has nothing to do with gaming, its structure reveals its heritage. There's a tutorial section (exposition, in the lingo of movies) where the characters can learn the rules, explore, and experiment safely; four distinct levels with enemies of increasing number and difficulty (including something akin to a cheat code, allowing the protagonists to move through one of the levels faster); and a recurring boss who the main character has to confront in the final level.[32] By chance or design, Nolan created a film that uses the framework and mechanics of video games to tell its story. He did so with mastery and elegance. And he scored a hit.

It seems likely that the initial concept of convergence—games begetting movies, movies begetting video games—was too narrow. Whether it's possible to create a game or movie adaptation that slips past the boundary and finds success on the other side may not be important. Or maybe game and movie studios were proceeding from the wrong assumptions. This last is an assessment that Joe Minton, CEO of DDM Agents, appears to share.

DDM Agents is essentially a management consultant to game developers, helping them get contracts, providing business advice, and building relationships between studios and publishers. Minton's been in the industry for more than two decades, first as head of a Massachusetts-based development studio called Cyberlore, then as a partner, and now CEO, of DDM. He sat down with me one morning at their offices in Northampton, Massachusetts, and shared some of his thoughts. "A lot of people want to make parallels between the movie business and the game business," he told me, "and I think the differences are greater than the parallels. At the end of the day, one of them is a linear, passive storytelling format, and one of them is an active, engaged, nonlinear reactive format, and there's many ways in which the development of those two medias just are done incredibly differently." It's the failure to appreciate that—particularly from the film side—that's led to most of the problems in effectively morphing them: Because the two industries are more dissimilar than they are alike, it's difficult—if not impossible—to create an entertainment property for one of them using the creative mindset of the other.

But you can share resources and techniques between them, and that's where convergence is really happening—fast.

• • •

I met Shane Culp in Boston in mid-2008. At the time, he was living in New York City, working as a software engineer and game designer for MTV Networks. We both happened to be at the same game conference, and after speaking briefly, we agreed that a follow-up conversation was in order. A few months later, I made my way to New York and we spent an afternoon talking. Shane is soft-spoken and has a slight accent that betrays his time in the Dallas area. One of his first jobs was as a game designer for casual game developer BlockDot, and while he was there, BlockDot did a promotion for *Star Wars: Revenge of the Sith.* This was in the late 1990s, early 2000s, near the beginning of the technological handshake between movies and games, when movies really started relying on digital assets. For the game Shane was creating, Lucasfilm provided him digital models from their own library. He remembers it well. "They actually were able to give me models of Obi-Wan Kenobi. We couldn't use them directly 'cause we were needing them at much lower resolution, but we could pre-render them and use them as assets. We actually used assets from Lucasfilm."

The reason they couldn't use the digital models directly is because the machines available to BlockDot for development—and ultimately the ones people would be using to play the game—weren't powerful enough to handle Lucasfilm's high-resolution digital models. "The film industry can use a much higher poly[gon] count because they're not doing real-time rendering,"[33] Shane explained. "They can take longer to render it out, so if it takes several minutes to render one frame, or hours, even, that's fine with them."

So, Shane had to lower the resolution of Lucasfilm's digital models in order to use them, but he *was* adapting actual movie assets for a video game. It was a start. At the same time, however, George Lucas had put a project into motion that would drive convergence significantly forward. A friend of Shane's was a lead engineer at LucasArts, and he gave him the details.

> What George Lucas did was he purchased a lease on the Presidio, and he went in and remodeled a big section of it, and he relocated his business. Lucasfilm and ILM and LucasArts, he decided to bring them all together at one location in the Presidio. He wanted his film industry guys right next to his game developers with big, giant Ethernet cables, and they wanted the ability to have all that digital content that's created either for a game or for a movie to be accessible and usable by both.

This being the mid-2000s, there were some technological hurdles to clear before Lucas' vision could be fully realized—the aforementioned trio of polygon count, image resolution, and computing power among them. Nonetheless, he'd set events inexorably in motion and laid the foundation upon which true convergence would be built. When Shane and I spoke in 2008, things were already changing. "What's weird is that we're getting to a time right now where this convergence is occurring," he said. "Really, when George Lucas was thinking about this and sort of building the Presidio, it probably wasn't real viable, but now, two or three years later, it actually is becoming very viable."

Viable indeed. The production teams that brought Paramount Pictures' *Iron Man* movie and Sega's accompanying video game to life shared assets, computers, and software between them, *Kung Fu Panda*'s effects people used a game engine for at least some of the film's animation, and Shane knew of other film industry animators who'd employed game engines for animatics and storyboarding.[34]

But this was only the beginning, the tip of the iceberg. It would take another four years, an announcement at the eighteenth Electronic Entertainment Expo (E3), and a conversation with one of ILM's own visual effects supervisors to appreciate the magnitude of change bearing down from just beyond the horizon.

ILM: THE EMPIRE STRIKES BACK

Being in ILM's lobby at the Presidio was one of my more surreal and awe-inspiring moments. There I was, standing across from Darth Vader and Boba Fett, two of the most notorious villains in science fiction, on hallowed ground. This was the studio that had produced the most memorable movies of my childhood, movies I'd spent countless hours emulating in various forms, and that I'd shared, at one time or another, with the most important people in my life. For a moment—just a moment—I forgot why I was there.

Right. Video games. Just seven weeks before, LucasArts had given the E3 crowd a glimpse of the future, and I was about to talk to Kim Libreri, ILM visual effects supervisor and part of the team behind the game poised to change the world.

We began in a demo room, in front of a massive flatscreen monitor—the kind that makes grown videophiles fall to their knees and weep—for a private viewing of LucasArts' newest addition to the *Star Wars* canon: *Star Wars 1313*, a grittier, more mature approach to *Star Wars* that takes place, in Libreri's words, in "the scariest place in the universe"—the seedy and crime-ridden Coruscant underworld.[35]

We settled in for the demo, and Libreri gave me a little background as it started up. The creative team behind *1313*, he said, "is genuinely a combination team between LucasArts and ILM. We had a really big chunk of ILM people join in to be really a part of the game. The animation that you'll see was animated by ILM animators, the rendering tech was put together by a combination of the ILM people and the LucasArts people. The facial capture system is the ILM facial capture . . . it's totally a convergence project."

He started the game, and the first thing that hit me—literally—was the sound. Spacecraft rumble overhead, shaking your seat. Explosions rattle the walls. Voices echo eerily inside your ship. Environment noises come at you with weight and presence. The score is heavy, foreboding. *1313*'s sound is physical, another character moving the story forward, leading you down a dark and dangerous path. If there was any doubt before, it has been erased: this will not be good.

Your ship descends below the surface, and the walls of Coruscant's innards rise up around you, imposing monuments to the gathering darkness. Lower and lower you drop, down into the bowels of the city, towards the game's namesake: level 1313, last stop on the Underworld Express.

I was utterly spellbound, captivated by the spectacle before me. The combination of lush imagery and deep, mood-altering audio knocked me into one of Seth Schiesel's moments of cognitive dissonance and held me there, completely enthralled.

And then the action stopped. There was still activity on the screen, but the main character wasn't moving—and after a moment, I realized why: We'd been watching a cut scene—a scripted cinematic device that games often employ to move a story forward and provide additional information beyond that discovered through gameplay. Cut scenes are, in essence, mini-movies. Because they don't typically run in real-time, they can use higher-resolution images and animation, and thus look very different from the actual game—so it's usually easy to tell when a cut scene is over. The problem here was that this cut scene ended, and Libreri hadn't noticed. He was supposed to be controlling his character, but he had gotten as caught up as I and had forgotten to start playing. There'd been no visual change between the animation and the game, and no pause to indicate the end of one and the beginning of the other. Libreri wasn't playing on some super-fast, top-secret machine, either. True, he was using a high-end PC with the latest Nvidia graphics card, but nothing that's beyond the access of even those of modest means (as of January 2013, a high-end gaming PC would run you around $1,500). And his team hadn't simply lowered the visual quality of the cut scenes or increased the graphic resolution of the game segments. What LucasArts achieved is nothing short of extraordinary: real-time rendering, at thirty frames per second, of near photo-realistic animation, and seamless transitions between cinematics and gameplay. No one had ever seen anything like this before. They couldn't have—because until now, it was impossible.

Part of it is the effect of Moore's Law. First described in 1965 by Intel co-founder Gordon E. Moore, Moore's Law simply states that the number of transistors you can fit on a computer chip doubles every two years. The practical result of this (and the technological reason *1313* can exist) is that computer power, processing speed, and storage capacity also double in nearly the same time.[36] For game designers, this means that the technical ability to realize their wildest visions is becoming a reality—and *1313* fits squarely into that category. Demo over, and comfortably installed in his office, Libreri explained it to me.

> We've got to a point where many of the techniques that we would have taken as our bread-and-butter at ILM a decade ago are now absolutely achievable in real-time. In fact, you know, a render at ILM—or any visual effects company—can take ten hours per single frame. This is running at thirty-three milliseconds—that's like a million times faster than what we would have on a normal movie.[37] But the graphics technology is so advanced now that the raw horsepower is there to do some incredible things.

Case in point: the demonstration I'd just witnessed. Though it's heavier on the cinematic component than the final game will be, it provides a window into the creative process behind *1313* and illustrates very clearly what skilled developers can now do. Libreri's team used professional television actors cast specifically for the main roles and then suited them up in gear identical to what they'd use for traditional performance capture: facial dots to pick up precise facial movements, stereo head camera setups, motion-capture body suits—the exact same system ILM employs when creating a feature film. They record action and audio at the same time, process the performances in a computer, lower the resolution a little bit—to about a quarter of what they'd use for a movie, but still far beyond anything previous—and then incorporate it into the game, adding elements, in real-time, like lighting and an ILM shading system that causes in-game objects to react to light the way they would in the physical world.

Star Wars 1313 owes more to its existence than technical wizardry, though. The heart of the game beats to a rhythm as ancient as humanity: storytelling. "The second part of it," said Libreri, "is having a creative director that really wants to immerse the audience and have them feel that they're part of a world and playing characters at a level of believability and integrity that you've never seen in a video game, so he really wanted the best in storytelling."

In any medium, the key to captivating an audience and transporting them into the story is making them feel down to their marrow that the world they're visiting, alien though it may be, is real. It's not realism—it's getting them to suspend disbelief in the face of the unfamiliar. When the medium is visual, the more realistic the world appears, the better the story has to be. If you put all your resources into creating a fully realized visual experience and give no thought to the story, then it's all for naught. Your world could be as elegant as the Taj Mahal or as lush as the Amazon basin, but without a believable narrative it would collapse under its own weight.

In video gaming's recent past, designers could get away with a mediocre story as long as the game looked good and played well. Now, Libreri believes that's all changing.

> You know, we've had the current generation of [gaming] hardware out for a while . . . audiences are on the edge of their seats for something new. In the same way as computer graphics revolutionized what you could experience in a movie theater, or what we did on the *Matrix* movie, it changed people's perception of what an action movie can be about. Gamers are looking for the same thing, and I think that technology is gonna sort of be the catalyst that enables it. But the big challenge is how do you design a game that works in that space? You know, the writing, if the writing . . . I downloaded a game that technologically was interesting, that came out earlier in the year, but straight away, I'm like, "God, the acting is really rough." The story was really cheesy and contrived, and I stopped playing it.

The problem is that there's too much competition and not enough time. A typical game can take upwards of forty hours to finish; given the choice between a good game with lackluster characters and story and a good game that also has a great story and a complete world, people will choose the whole experience—and this is starting to push convergence on another level. Video games have reached the point where you not only need great designers, artists, animators, and technical people, you also need great directors, writers, and actors—the heart and soul of Hollywood.

"Think about a mocap shoot like that," said Libreri. "If you're gonna do a large component of cinematic content, you've got actors that you're having to cast differently. No longer can you use a guy who's got a great voice but can't act. I'm sure there'll always be work for voiceover actors, but if you want a leading actor, then you're gonna want a leading actor."

Actors, after all, are the ones ultimately responsible for telling the story. Yes, there are techniques that filmmakers use to guide audiences through a movie and help them understand what they're seeing and what the filmmaker is trying to tell them—lighting, shot composition, performance delivery—and Libreri feels that video games will have to adopt these same techniques and ideas going forward. The best games, he believes, will be those that think carefully about where players are going to look, where they'll get stuck, and how they'll move from one level to the next—in short, the ones that are really smart about how they convey their stories. It's the actors, though, who are the face of a game, who establish the emotional connection, and who provide the link between the player in front of the screen and the world behind it.

We talked for a bit longer, then I left Libreri's office and wound my way back to the main lobby—stopping first at Java the Hutt for a hot mocha (I couldn't resist). On the way, I passed icons from my youth: production drawings from the *Star Wars* films, models of Darth Vader, R2-D2, and Chewbacca, a few of Ralph McQuarrie's original matte paintings. It was hard not to feel like a kid. I thought about the time I'd spent talking with Libreri. In another five to ten years, he'd said, video game graphics would achieve absolute photo-realism. Computer processing speed, power, storage capacity, Internet bandwidth—all these hurdles to completely immersive entertainment and true virtual environments would be gone. From that point on, the world would be all but unrecognizable.

Yet we, as human beings, would be fundamentally the same. We'd still have our anxieties and fears, our hopes and dreams. And we'd still have the need to play and the desire to be transported from our daily lives to a grander, if temporary, existence. I was reminded of something else Libreri had said. "Gameplay," he told me, "is not just about your skills and reactions. It's about what your brain and heart perceive when you're playing the thing, and most people will choose to be immersed in a world that is vibrant and interesting, and makes them feel that they're actually part of something bigger than just pressing a button and killing an alien."

Ultimately, I think that's what we're all looking for: To feel part of something larger than ourselves, to do great and heroic deeds, to triumph against impossible odds, save the day and ride off, victorious, into the sunset. I found that in the movies of my youth; video games serve the same purpose for the youth of today. For all the talk about the merging of film and video game, and for all its inevitability, perhaps the secret of true convergence lies not in an external reality, but in an internal truth: What kids seek from video games is what we all seek from our own distractions—be they movies, radio, comic books, literature, or art: an escape from the mundane to the sublime, where our imaginations make of us heroes, lovers, warriors, and gods.

8

VIRTUAL LIFE

Laura Skye and Dave Barmy met in 2003 in an Internet chat room and became fast friends. Their online friendship quickly blossomed into something more, and the pair decided that they wanted to be together, so Laura made the leap, leaving her home and moving to his. They spent almost all their time in each other's company and would often be seen out on the town together—she in skin-tight cowgirl outfits that accentuated her slender figure, he sporting long hair and sharp suits over his well-muscled frame. Laura and Dave had similar interests, and they enjoyed one another. They were happy.

Or so it seemed.

One day not long after they'd moved in together, the unthinkable happened: Laura caught Dave in bed with another woman, a prostitute. Shocked and horrified, Laura broke off the relationship. But she was still in love—so she decided to test him. In desperation, she turned to a private eye—a woman named Markie Macdonald, who set up a honey trap for Dave, employing an attractive coquette to approach him at a party and seduce him. The trap failed, though; Dave remained faithful to his love, talking about Laura all night. Laura forgave him, and the two rekindled their relationship.

In 2005, Laura and Dave were married in a ceremony in a beautiful and exotic tropical grove. For a while, all was well. A few years later, though, Laura began to sense trouble. In April of 2008, her suspicions were confirmed: she caught Dave in another woman's affectionate embrace. Unlike his previous indiscretion, there was no sex. This time, it was worse; the two seemed genuinely in love—and when Laura confronted Dave, he admitted as much. He had deep feelings for this other woman and was no longer in love with Laura.

She knew it was the end; there was no hope of reconciliation this time. That same month, Laura filed for divorce, and in May of 2008, after a short, intense relationship, Laura and Dave's marriage was over.

• • •

Not exactly surprising, is it? People fall in love and marry—advisedly or not—for many reasons. And they divorce for as many, if not more. The difference here is that, although this is a true account, it never happened—at least not in the way most would imagine. Laura Skye and Dave Barmy are avatars—virtual representations of actual people,[1] in this case twenty-eight–year–old Amy Taylor and forty-year-old David Pollard—that exist only within the virtual reality of *Second Life*.

If you're unfamiliar with it, *Second Life* is a virtual world[2] developed and owned by San Francisco–based Linden Labs. Launched in 2003, *Second Life* provides users (residents, in the virtual lexicon) a space to build objects and landscapes, run businesses, trade property and services, participate in events, visit attractions, or simply interact with others in-world. The scope of what you can create is limited only by your imagination: There are businesses that sell virtual clothing for *Second Life* avatars, recreations of global landmarks, dance clubs, shopping malls, areas dedicated to conservation, adult-themed hotspots, art museums, and science centers. Real-world organizations like IBM, the Centers for Disease Control and Prevention, NASA, and the National Oceanic and Atmospheric Administration have even set up branches in *Second Life*. And then there are the events. Through their avatars, residents can attend college lectures, political rallies, theatrical and musical performances, exhibitions, and readings from anywhere in the world: A student in China can hear a presentation from a university professor in France or a music fan in Australia can catch a concert from her favorite US band without leaving their homes. For people without the means to travel, *Second Life* allows them to experience events or visit places that are otherwise beyond their reach.

For many, though, *Second Life*'s appeal lies in two aspects: the freedom to be whoever they want and the ability to connect and form relationships with real people. When you create a *Second Life* account, you also create your own avatar, in whatever image you choose. It can be taller, shorter, thinner, or more muscular or athletic than your own body. It can have more hair, less hair, no hair. You can follow your own gender or switch to the opposite (a fairly common practice, actually). You don't even have to be human: there's a group called the Furries, who are dedicated to, as author Julian Dibbell puts it, "role-playing the lives . . . of cuddly, anthropomorphic woodland creatures."[3] Once you've designed your digital self, you can then explore the world of *Second Life*. Spend enough time in it, and you'll inevitably meet people who you'll become friendly with. Perhaps there will be one who you develop a particular fondness for—even an attraction; if it's reciprocated, the two of you may fall into a relationship, move in together, and get married.

And that's where it can get complicated. You see, you probably have no idea who the actual person behind the avatar is—gender, race, sexual preference, marital status in the real world . . . it's all a mystery. You may even be in a relationship outside *Second Life* yourself. One thing's for certain, though: the feelings you have are unquestionably genuine. This isn't necessarily an issue; many people find it easy to maintain a divide between worlds. But many don't, and that's when the virtual can become painfully real. Return with me to the couple we opened with, Amy and David (and their avatars, Laura and Dave), and you'll see what I mean.

Okay, confession: the romance between Laura and Dave didn't happen *exactly* as I presented it: the avatars didn't meet in 2003, Amy and David did (though still in an Internet chat room), and it was Amy who moved, not Laura. The rest is accurate—and it had some serious real-world repercussions. When Laura and Dave married in that lush, tropical setting in *Second Life*, Amy and David were also married at a distinctly less exotic register office in Cornwall, England. And when Dave transgressed his matrimonial oath to Laura three years later, it also doomed David and Amy's marriage: the avatars divorced in May, Amy and David followed in November.[4] That the infidelity occurred strictly within the virtual world was irrelevant. To Amy's mind (and, more importantly, her feelings), her husband was involved in an emotional relationship with the woman behind the avatar. "It may have started online," she said, "but it existed entirely in the real world and it hurts just as much. His was the ultimate betrayal. He had been lying to me."[5]

Think about that for a moment. An actual marriage ruined by an affair that took place in a virtual world, between the avatars of two people who never touched, never met, never set eyes on each other—hell, were never even in the same country. None of that mattered. Perhaps Amy should have been able to get past it. It wasn't "real," after all. But she couldn't—and she's hardly alone. Suffolk University's Nina Huntemann has been doing focus groups with so-called "game widows"—women who, to some degree or another have been abandoned by their partners for the thrill of a game—and their stories are heartbreaking. "The way they describe betrayal and neglect and abandonment and cheating from a partner who has developed a relationship with another woman online," she said, "it sounds to me as real, and their pain is as real, as any relationship in the physical world. At some level, emotionally, it's the same betrayal."

Tales of the virtual world intruding on the real abound, some with tragic results—as in the 2005 case of a Chinese man stabbed to death over the theft and sale of virtual sword from the fantasy world of *Legends of Mir*.[6] Granted, that's the exception. Most cases of virtual theft don't end in actual murder—but they're not without other consequences. In 2007, a Dutch teenager ran afoul of the law when he stole close to six thousand dollars worth of virtual furniture from users of *Habbo Hotel*.[7] And there's my personal favorite: the 2008 arrest of a Japanese piano teacher accused of murdering her virtual ex. It seems that the woman's online husband (they were devotees of the South Korean *MapleStory*) divorced her without warning, to which she took less than kindly. In retribution, she logged onto his account and killed his avatar. The man reported the virtual foul play to the police, and she was taken into custody and charged with illegal access to a computer and manipulating electronic data. She faced up to five years in jail or a five thousand dollar fine.[8]

These examples may sound silly, the injured parties' responses extreme, but it illuminates a large and very significant issue: people often develop strong emotional attachments to their avatars, and events that happen to those avatars—good or bad, directly or indirectly—can affect them deeply. It's a phenomenon that the Oxford Internet Institute's Ellen Helsper is very familiar with. "For a while," she said, "there was this impression that as long as it's online, it doesn't matter. But research has shown it's not a separate world." Thus, infidelity is "just as painful, whether it's electronic or physical."[9]

For his book, *Coming of Age in Second Life*, anthropologist and author Tom Boellstorff talked to several *Second Life* residents who had broken up with or been left by their in-world partners.[10] They spoke of feeling a sense of loss as intense as at the dissolution of a real-world relationship—if not more so. Certainly, the heartsick behaved like it, avoiding people or places associated with their ex-partners, refusing to wear items of virtual clothing or view in-world snapshots because of the painful memories they evoked.[11]

But if virtual infidelity and betrayal can be felt acutely by its real-world victims, so too can virtual friendship and love. Bear in mind, before Amy and David's marriage fell apart, they were, though Laura and Dave, very much in love—as much as any bygone paramours known to each other solely through sensual missives carried surreptitiously in the dark. How could Amy have taken David's *Second Life* affair so firmly to heart otherwise? Why, then, shouldn't two people be able to *successfully* navigate the sometimes rocky waters of virtual intimacy?

As it turns out, they can. Boellstorff witnessed this on several occasions. One couple he met in-world had been married within *Second Life* for three years and had a "close and meaningful" relationship despite never having shared so much as a photograph in the real world.[12] For many, being real for each other within the confines of *Second Life* was enough. "The virtual romance was complete within its online parameters," he wrote. "It was real, because the intimacy, care, and desire were real."[13] A *Second Life* resident from Europe related a story about her in-world partner, who lived in the United States, sending her a website that had a live camera feed from a highway near his home. The two snuggled on their virtual couch, in their virtual condo, and watched the real sunset together[14]—an activity that in the physical world, though possible, would have required a substantial investment of both time and money.

Some of the residents Boellstorff interviewed actually felt that a virtual romance was more authentic than any real-world liaison could ever be. *Second Life*, they observed, allowed you to get to know people beyond physical appearance, free of any preconceived ideas or prejudices you might have if you encountered them in "real" life. Occasionally, *Second Life* even allowed for expressions of intimacy that would be difficult, if not impossible, under any other circumstance: He knew of two women in-world who maintained a lesbian relationship for months—even after learning they were both, in reality, heterosexual men.[15]

Opportunities for love aside, there are times when someone just needs a sympathetic embrace—a situation Boellstorff experienced first-hand. He recalled a particular day when he'd been trading messages in-world with his friend Vonda, and could tell from the tenor of her responses that she was down and out. She really needed a friend, she said, and messaging wasn't enough; she wanted to be held. Could he come over?

Boellstorff did what any good friend would: he immediately teleported to Vonda's location (one inarguable fact: *Second Life*'s got the real world beat when it comes to getting around), found her on the first floor of her house, and held her. Then, arms wrapped around each others' shoulders, they began to slow dance. "I can feel you," she typed. "You are a nice man. I've been having a hard time, and I need time to relax with a friend. Hold me tight, don't let go. I need friends now."[16]

This is exceedingly odd. Boellstorff and Vonda had been chatting in *Second Life* via instant message, and she asked him to come over and hold her. Not his avatar, him. So he did, and when he arrived, he enfolded her in a virtual embrace. There are two items of note here: first, there's the language, the lack of separation between the flesh-and-blood Boellstorff and his digital manifestation.[17] Second—and far more fascinating—is the interaction itself. Physically, there was absolutely no difference between the messages and the "in-person" communication: Whether you're instant messaging back and forth or speaking directly, avatar to avatar, it's all text entered by keyboard; *Second Life* makes no distinction between the two. And yet, it made a difference to Vonda. She said she could feel him, and it *helped*. That virtual hug made Vonda feel better, safer, protected perhaps. At the very least, she felt like someone cared, like she had a true friend.

Okay, what's going on here? Why did this work?

Three words: the Proteus Effect.

First described in 2007 by Palo Alto Research Scientist Nick Yee (who coined the term), the Proteus Effect relates to the ability of an avatar's physical characteristics to alter how a person behaves within a virtual world.[18] In his initial research, Yee provided study subjects with avatars that were attractive or unattractive, tall or short, and then watched them interact with a virtual stranger (controlled by one of Yee's lab assistants). He found that participants who had attractive avatars walked closer to the stranger and revealed more personal information than those with unattractive ones. With respect to avatar height, those inhabiting taller avatars (by ten centimeters, compared to the lab assistants) were more aggressive negotiators than those whose avatars were shorter, and were also much more insistent on getting the better end of the deal. Yee's work demonstrated clearly that an avatar's appearance could change how someone acted within a virtual environment and interacted with its residents.[19]

A crucial question remained unanswered, however: Did these changes in virtual-world behavior translate back into the real? To shed some light on this, in 2009 Yee revisited his study, conducting the first part of the test as mentioned, but adding another task: After concluding their virtual interaction, Yee had each participant create a personal profile on a mock dating site and then, from a group of nine possible matches, select the two s/he'd most like to get to know. Without fail, subjects who'd been given attractive avatars in the first task chose more attractive people to meet from the online dating site in the second—and the opposite was true as well. For the first time, Yee had shown that a person's sense of his or her own attractiveness in the virtual world could actually affect their real-world sense of self, at least temporarily.[20]

Yee's research illustrated the Proteus Effect at work within limited confines. There are other individuals and organizations who are looking at it in a broader social context. Maria Korolov, president of Trombly International and founder and editor of the online publication *Hypergrid Business* (which covers business use of virtual worlds), related a personal experience to me one afternoon. "I was touring the *Second Life* demo grid—the construction guy who built it gave me a private tour—and there's a street. I was crossing the street, and I looked both ways." She paused and laughed. "I'm an avatar. A, it's an empty grid, no cars coming. B, even if they were coming, they're cartoon cars. But I'm still looking both ways."

Now, Korolov's a very intelligent woman, and she's studied immersive worlds from their inception, so she's no neophyte. And yet she still had an automatic, unconscious connection to her avatar—so much so that she brought life lessons with her into the virtual world. And she's not alone: studies of social interaction between avatars reveal that avatars keep the same distance between each other as their human counterparts do in the real world—and that distance changes, depending on the culture and background of their creators.[21]

As Yee discovered, the exchange works both ways. "If people exercise in the virtual world," Korolov told me, "they will exercise an hour more on average the next day in real life, because they think of themselves as an exercising-type person. It changes the way you think. People who are taller avatars are more aggressive in meetings in the virtual world, but . . . they continue to be more aggressive in meetings and negotiations in the physical world for a few hours afterwards."

The larger point, according to Korolov, is this: If your sense of connection to your avatar is strong enough—if it looks like you and moves when you move it, for example—you can identify with it to the point where something that happens to the avatar in-world actually, physically affects you.

Which is exactly what happened with Vonda: so profound was the link between avatar and woman that Boellstorff's virtual touch provided solace to her physical being.

If that seems bizarre (and I'll freely admit that it is), then hold on to your hat, because it gets weirder. As humans, our primary experience of the world is visual: a good bit of our brain is dedicated to transmitting, processing, filtering, and interpreting images gathered by our eyes. And *Second Life*, like almost all virtual worlds, is primarily a visual medium—so though it requires some cerebral acrobatics to absorb the avatar/body connection, it is perhaps not unreasonable to think that an intensely visual experience would have such an effect on a predominantly visual species. Here's the thing, though: those gorgeous images, the lush landscape, that perfectly designed avatar . . . completely unnecessary. All you really need is an immersive world, a group of people to interact with, and a healthy dose of imagination. Of course, it *can* be as involved and graphically rich as *Second Life*, but for the purposes of creating the virtual/real connection, a text-only multiuser dungeon will do the trick.

Consider a strange and disturbing event that occurred about twenty years ago in *LambdaMOO*.[22] Founded in late 1990/early 1991, *LambdaMOO* is a text-based virtual world that allows multiple users to connect simultaneously, explore their surroundings, and interact with each other. Navigation and conversation take place by way of typed commands or dialog. Users create their own avatars by entering descriptive text into the system, which is stored within *LambdaMOO* and retrieved each time a user logs in. One of its physical features is a living room—a large common area where much of the socializing takes place. It was here where the incident in question unfolded. A new character—Mr. Bungle, he was called—appeared in the living room one evening when it was packed with established players, most of whom had known each other for some time. By way of introduction, he started up a "voodoo doll" (a small program that gave its user control over other avatars) and commanded one of the players—a somewhat androgynous female character named legba[23]—to service him in a variety of sexual ways, against the player's will and completely beyond her control. He then forced another avatar to perform sexual acts with others in the room and then brutally violate herself with a kitchen knife, before he was finally stopped. The *LambdaMOO* community was outraged and called for various actions to be taken against Mr. Bungle, including his virtual execution (permanent removal from *LambdaMOO* by its administrators, which is ultimately what occurred)—all the while rallying support around the victims of his attack. Most fascinating, though, was legba's reaction, which was, in part, "I'm not calling for policies, trials, or better jails. I'm not sure what I'm calling for. Virtual castration, if I could manage it. Mostly, [this type of incident] doesn't happen here. Mostly, perhaps I thought it wouldn't happen to me. Mostly, I trust people to conduct themselves with some veneer of civility. Mostly, I want his ass."[24]

The more sobering aspect of her reaction, though, is this: While she was venting her spleen in the virtual world, she was crying tears of post-traumatic stress in the real.[25] That the attack—the rape, if you will—happened strictly within a text-based, virtual realm had little bearing. It was still a violation: she'd lost control of her avatar, who was forced to engage in sexual acts against her will. And even though *LambdaMOO*'s representation was entirely textual, with none of *Second Life*'s visual component, the bond between legba and her creator ran so deep that the vicious transgression against legba produced acute trauma, anxiety, and stress in the woman who brought her to life.

This is, perhaps, one of the more extreme instances of the Proteus Effect in action. Regardless of scope of impact, though, the Effect carries with it some implications that are astounding and, quite frankly, of real concern. If a virtual event within an entirely text-based world can so powerfully affect someone, or if, as with Yee's research, simply dictating certain features of an avatar can impact a person's feelings of value and self-worth that reach out from the virtual world to the real, then who knows what you can make them believe. Are there any limits?

It's a point taken up by Jesse Fox, Jeremy Bailenson, and Liz Tricase at Stanford University, in research that further extended Yee's work. Their study explored the highly sexual representation of women in avatar-based virtual worlds and video games, and what they learned was disturbing.

Participants in the study (all women) entered a virtual space as one of four possible female avatars—two dressed provocatively (tight clothes, short skirts, long hair, more skin visible) and two conservatively (long pants, loose-fitting shirts and jackets, hair pulled back or up). The researchers had taken digital photographs of each subject beforehand; as an added dimension, they mapped those photos onto the heads of the avatars so that individual participants had avatars customized to look like them. While immersed in virtual reality, they engaged in a series of actions, including basic movements in front of a virtual mirror (to establish connection with the avatar), conversations with male avatars (controlled by research assistants), and simple tasks assigned to them by the male avatar. Following the session, participants completed a short questionnaire, and that was that.[26]

After reviewing the results, the researchers found evidence of the Proteus Effect at work. Inhabiting sexualized avatars within the virtual space had psychological effects that carried over into real life: Women in this group reported more body-related thoughts (weight, size, and appearance, for instance) than those in the group of more reserved avatars. That's not much of a stretch; it jives with research conducted in 2006 about women trying on swimsuits versus sweaters[27] (guess which group was preoccupied with body image). The worry is that women who choose to represent themselves in virtual worlds through sexualized avatars may perceive themselves as objects—good for sex, but not much else. Even worse is the issue of rape myth acceptance—the belief that women who are raped somehow deserve it, that they brought it on themselves: They wear provocative clothes, drink too much and lead men on, stay out late, etc. There's a fear that women who create and/or use suggestively or overtly sexual avatars may begin to incorporate this idea of rape myth acceptance—that, in essence, they get what's coming to them—into their psyches, and therefore validate it—and that men who encounter these avatars (or use them themselves) will do the same. "In this study, simply wearing certain avatars led women to blame rape victims for their assault," the authors wrote, adding that this was "a dangerous attitude for a woman to have as a potential juror, confidante, voter, family member, or even a victim herself."[28]

At this point, the jury is out: The study of virtual worlds is still a very new pursuit, and much more needs to be done to determine the validity of these fears. Nevertheless, as we continue to travel further into this brave, new virtual world, we must do so with caution, and tread lightly until we're sure of the ground beneath our feet.

Of course, not all is to be feared. Virtual worlds may indeed present us with unique challenges and unknown perils, but they also have the power to change—and even save—lives.

Allow me to introduce you to Jani Myriam. Jani is a twenty-something young woman, attractive, spunky, confident . . . and entirely virtual. Jani is the alter ego of William Wise. He created her in 2006 to allow him to experience life as a woman—something he'd always wanted to explore but found impossible . . . in the real world, that is.

And then *Second Life* changed everything. For the first time, Wise could try living as a woman and take the initial steps towards creating the life he felt he should have been born with. "I liked myself so much better as Jani," he said. "She was fun, happy, even bold and witty, while the real-life me was overwhelmed with fear and self-doubt."[29]

Life as Jani was everything that life as William wasn't, and he loved it. Finally, he was fulfilled. After living within *Second Life* as Jani for close to two years, he came out as a transsexual and began preparing for the operations that would transform him from William to Rebecca in the real world. Without the ability to safely try on his new identity in the virtual world of *Second Life*, Wise may never have had the courage to change genders—forced by circumstance and biology to spend his days trapped in the wrong body.

Wise went on to found the Transgender Resource Center (TRC), a sanctuary within *Second Life* where others like him could go for counseling, answers, and support. For transgendered people, this was a critical resource: According to Melady Preece, a clinical psychologist at the University of British Columbia, they're much more at risk of suicide than the general population. "I have no doubt," she said, "that the volunteers at the TRC have prevented, whether knowingly or unknowingly, a number of suicide attempts."[30]

Second Life provides safe haven for more than just transgendered folks, though. People with a range of concerns—alcoholism and drug use, cancer (survivors as well as those still living with the disease), stroke, autism, post-traumatic stress disorder, mental health issues—can find support groups in-world. And because you can create an avatar in any image, you can be as anonymous or open as you like—revealing or concealing as much of yourself as is comfortable.

Perhaps nothing is as poignant, though, as the stories of the severely disabled walking, running, even flying through a world where their physical limitations have no bearing, taking part in activities that, in real life, may forever be out of reach. Watching a wheelchair-bound, nonverbal child log in to his favorite world and come alive in a manner that is otherwise impossible is at once inspiring, breathtaking, and heartrending. For these people, virtual worlds like *Second Life* aren't merely pleasurable distractions, they're portals through which they can escape from their broken bodies and simply live, unbounded, in ways that most of us take for granted.

For all the talk of virtual worlds being bizarre and different—and for all the ways they actually are—it's in their resemblance to the real that they're most fascinating, for therein lies their true power. Yes, people choose different identities, reinvent themselves, play out fantasy lives, switch genders, and explore alternative lifestyles. Many experiment in areas they would never dream of outside the virtual, and in so doing, may learn something of themselves. But the vast majority of what happens there is simply human drama, played out on a grand stage of our own design. Virtual worlds are both a mirror reflecting the virtues and ills of our society and a gateway through which we can realize our true potential and fully express the essential nature of our humanity.

9

. . . AND WE ARE MERELY PLAYERS

Video Games and Society

September 13, 2005. Tuesday. Sunny, humid, an occasional breeze shifted the heavy air. Temperatures averaged in the mid-seventies, though here in Massachusetts, the mercury would climb to a record-setting eighty-nine degrees. In New York, voters headed to the polls for the state Democratic primaries, while on the national stage, the Senate Judiciary Committee opened confirmation hearings for Chief Justice nominee John Roberts. On balance, it was an unremarkable day for the world to end.

A group of explorers picked up the pathogen in a remote corner of the jungle. Previously unknown, highly virulent, and extremely contagious, it quickly infected the entire party. Had they all died there, had that first contact wiped them all out, everything might have been fine. But they didn't. Riddled with disease, they found their way out of the jungle and back to the nearest city.

Hindsight, as they say, is twenty/twenty. If those first victims had known the hell they were about to unleash, they might have stayed in the jungle, sacrificing themselves for the greater good—or, at the very least, undergone voluntary isolation until help arrived. But that's not how it went down. The explorers came into the city through one of its main ports of entry; one could scarcely imagine a more serious mistake. Anyone exposed to them quickly fell ill and then served as secondary sources of infection, transmitting the disease to everyone they came into contact with. A few cases became dozens, then hundreds, then thousands. And they weren't just getting sick.

They were dying.

Horribly.

The virus was like something out of a nightmare: Its victims erupted torrents of blood, spattering pestilent fluid over anyone unfortunate enough to be nearby. Panicked, delirious, the infected ran while they could—from each other, from themselves, as if they could somehow escape the horror consuming them.

The disease spread like wildfire. Transmission and mortality both approached 100 percent: if you were exposed, it was a safe bet you'd catch it. And if you caught it, chances were good that you'd die. It was the perfect plague: it killed indiscriminately, taking the young, middle-aged, and old alike without pity. To make matters worse, the virus jumped species with frightening speed, crossing unhindered between humans and animals, finding additional reservoirs in livestock and pets.

Authorities had no idea what was going on; those in charge could barely identify it, let alone halt its progress. They attempted to isolate the disease, establishing quarantine zones throughout heavily populated areas, but they were completely overwhelmed, and the virus was spreading too quickly. Aid workers and volunteers who rushed to help often fell ill and died themselves. There were also those for whom curiosity outweighed fear: many risked death, entering hot zones or areas under quarantine just to catch a glimpse of the virus in action. Most never made it out. As the disease took its toll, the stricken piled up, victims dying too fast to count. Centers of commerce shut down as thriving cities became deserted wastelands, streets littered with the bodies of the dead. Many tried to run, but it was only a matter of time before they became casualties as well. The virus was mindlessly relentless and virtually uncontainable; there was little you could do but wait for your turn to die and pray that it would be quick.

This was the nightmare scenario: a new, highly contagious pathogen that killed without bias, crossed species effortlessly, and was all but unstoppable—an epidemiologist's worst fear brought shockingly to life. It became known as Corrupted Blood, and it was poised to become the worst pandemic in human history.

• • •

If you're reading this now, then we survived. Corrupted Blood did not become the world-ending pandemic everyone feared—though you might be wondering why global news wires weren't electric with reports of the disease, and why this is the first time you're hearing about it. Given the recent hysteria over H1N1 (you remember the swine flu panic, right?), you'd think that they'd cover nothing else. You might also be wondering how we finally stopped the virus, and if, Heaven forfend, there's a chance it could return, perhaps with a vengeance. Bad news first: yes, it, or something like it, could show up again—and this time, even more people would be at risk (in 2005, there were about six million, today it's closer to ten million). The good news is it doesn't matter, at least not for most of us. That's because Corrupted Blood was a virtual epidemic, confined entirely to the fantasy universe of Blizzard Entertainment's massively multiplayer online game (MMO) *World of Warcraft* (*WoW*).

WoW, you may recall, is a *Dungeons & Dragons*–inspired MMO wherein you create a character (your avatar) and wander through a gorgeous 3D landscape, undertaking quests and adventures, joining organized groups of like-minded players (guilds), or just exploring the vast and expansive world. Your avatar's characteristics—strength, knowledge, agility, etc.—are all determined by how many hours you've put into the game: the longer you play, the stronger and more adept your character, and the more you're able to do. A powerful, high-level character often represents hundreds or thousands of hours of real time spent in gaining experience and leveling up. Those gamers who reach the upper echelons can find themselves hitting the virtual glass ceiling, having killed the hardest creatures or completed the most challenging quests. To keep these long-term gamers interested and playing, Blizzard periodically releases updates to *WoW* that include, among other things, new and especially difficult monsters to battle. It was one such update—the introduction of the Zul'Gurub dungeon, home to the immensely powerful Hakkar the Soulflayer—that led to the escape of Corrupted Blood, a parting gift from Hakkar to those players who managed to kill him or at least bring him near death. Blizzard designed the pathogen to inflict a large amount of damage continually over time and also to be transmittable from player to player. The thought was that players would either kill Hakkar or die in the attempt, but that they would stay within Zul'Gurub until one of those two events happened. What Blizzard didn't count on was that a party of gamers might engage Hakkar in battle, become infected with Corrupted Blood, and decide to leave before finishing the job by teleporting out—and that's when the problems started. Though high-level characters could sometimes withstand the damage from the plague long enough to find a healer, it would finish off lower-level characters almost instantly. They'd catch the disease from the ones who brought it back—typically staying alive just long enough to infect those around them—and then start dying in droves.

Two additional factors aggravated the situation: nonplayer characters (NPCs) and battle pets. *World of Warcraft* is more than a game; it's a virtual world in its own right. In addition to its many game-based locations and events, there are also towns and cities where players can purchase supplies and other needed items from local shopkeepers—represented not by other gamers, but by game-controlled NPCs. In the early days of *WoW*, some players took great pleasure in repeatedly killing the NPCs—more of an annoyance than anything. Still, it affected game play, so Blizzard made NPCs nearly impossible to kill.[1] This had the unintended consequence of providing them complete immunity to Corrupted Blood—but they could still carry it. Like virtual Typhoid Marys, NPCs transmitted the disease to everyone around them, and as they never died, they became constant sources of re-infection.

The other complication was battle pets. Battle pets are animals who can be summoned to assist a character during combat. They can both inflict and receive damage, and as such were just as likely to be affected by Corrupted Blood as the player who beckoned them. They can also be sent back to the aether, and therein lay the dilemma: Dismissing a diseased battle pet stopped the clock on the plague, but did not cure it. When re-summoned, the clock started up again, the battle pet began taking more damage, and it then served as a vector for infecting others. If a player sent for his or her pet in the middle of a crowded city, everyone who came into contact with it would catch the disease, and the cycle would begin again.

What started out as a challenge to high-level players quickly snowballed into a world-changing event. Blizzard's software engineers fought for days to contain Corrupted Blood, but they were spectacularly unsuccessful. They eventually conceded defeat, shutting down and restarting three servers and resetting the game—and all avatars affected by the disease—back to its state prior to the unveiling of Zul'Gurub.

I first encountered Corrupted Blood in the context of a keynote presentation given by Dr. Nina Fefferman at the fifth annual Games for Health conference in Boston, Massachusetts. Dr. Fefferman is currently Associate Professor in both the Department of Ecology, Evolution and Natural Resources and The Center for Discrete Mathematics and Theoretical Computer Sciences at Rutgers University. She's Principal Investigator of the eponymous Fefferman Lab, where she focuses on "mathematical and computational models of biological systems related to epidemiology, evolutionary and behavioral ecology and conservation biology."[2] She's also on the faculty of Tufts University School of Medicine, as Assistant Research Professor of Public Health and Family Medicine and co-Director of the university's Initiative for the Forecasting and Modeling of Infectious Disease. And she's consulted for and/or worked with the Defense Advanced Research Projects Agency, National Defense University, and the Departments of Defense and Homeland Security.

Okay, I know that was a lengthy introduction, but it's important to know who you're dealing with here, because she's also a gamer—and an avid one—and is convinced that games like *World of Warcraft* can provide valuable insight into human behavior in the face of some pretty horrible events and can, in turn, inform national policy in preparing for and responding to them. If that sounds far-fetched, re-read the previous paragraph. She knows whereof she speaks, and if she believes that games have something positive—and perhaps critical—to offer, we'd do well to listen.

So I did. A few months after Games for Health, Nina and I met at her office on the Rutgers campus. It's bright and comfortable, and has various nods to geek culture scattered throughout—notably a fuzzy tribble that vibrates when you catch it, and a few giant, stuffed microbes (she pointed out *E. coli* and sleeping sickness). And she has tea, which she offers freely. Intentionally or otherwise, everything in the space has the effect of putting one at ease, which belies the seriousness of her research—as does her manner: if you ever had to hear really bad news, Nina's who you'd want to deliver it. She has a way of speaking about global pandemic and the end of the world with a nonchalance that suggests she's talking about bad weather spoiling your weekend cookout. She is charming, she is brilliant, and talking to her is simultaneously thoroughly enjoyable and one of the most frightening things you'll ever do.[3]

Nina's exposure to Corrupted Blood was a matter of timing and luck. She and one of her grad students—Eric Lofgren—were both devotees of *WoW*, and they had been joking about developing a disease they could model through something called the Zombie simulator—"a very cute, *Pac-Man*–like simulator of zombies," she told me. Three weeks later, Corrupted Blood hit. Eric noticed it first, having logged on to the game more recently, and immediately called her. She logged on herself and followed it through to the end.

I asked her about how tracking the progression of a virtual disease could help model a real-world outbreak. It wasn't so much the process, she said, as the content. "What we're looking at with Corrupted Blood is not actually new ways to model disease," she continued, "it's new insights into what to put into those models. So the techniques of modeling haven't changed as a result, it's more what to capture in terms of the dynamics. It's really a question of understanding the human aspect."[4]

As a disease, Corrupted Blood, though extreme, fit the classical model very well: It started in a remote location and spread by means of infected players teleporting to population centers (much like hopping an international flight from Nairobi to New York). It passed from person to person through proximity and contact with bodily fluid. Animals provided a second source of transmission, as the virus readily crossed species—similar to bird flu. There were asymptomatic carriers who remained healthy but could still pass the disease along. And players with the ability to heal others traveled to help the sick and dying, like aid workers dispatched to assist during an actual epidemic—some succumbing to the illness themselves. Everything fit, it all worked perfectly . . . except for one small wrinkle: human behavior. It's a frustration shared historically by epidemiologists all over the world. Human beings are hard to predict, and our actions are impossible to model. "As epidemiological modelers," Nina told me, "we're constantly trying to think like, 'okay, how can human behavior accidentally break our system?' Not because the math was wrong, but because we failed to incorporate the capacity for some behavior that would totally invalidate our outcome."[5]

This is where games like *World of Warcraft* can be instructive on a societal level. During the course of the Corrupted Blood pandemic, two behaviors emerged that caught epidemiologists off-guard. Nina calls them "curiosity" and "stupidity."

Consider quarantines for a moment. Quarantines are designed to break the cycle of continued transmission and stop a pathogen in its tracks. There are a few well-documented issues with quarantines: First, how long do you keep people in? Second, what happens if a person who's been quarantined for a period of time and has been symptom-free is suddenly exposed to someone who's ill? Does the quarantine clock start over (epidemiologically, the answer is a resounding "yes")? And third, what if people refuse to stay? These are factors that have long been taken into account. But what if someone tries to break *into* a quarantined area? Certainly, there are those who have legitimate reasons for being there: health care workers, researchers, epidemiologists, even journalists (though their entry and exit has to be handled very carefully). People who simply want to see what's going on and then leave again (curiosity)—or worse, who do it on a dare, perhaps challenging one another to break in and out without being caught (stupidity)—are another matter entirely. Clearly, once they get in, it would be best to keep them there. Better still would be to prevent them from entering in the first place—but that changes how you handle security. Traditionally, security around quarantines is one-way, aimed at keeping people in who don't want to be there. Looking bidirectionally requires an entirely different approach. As an example, Nina cited prisons. "No one's worried about people trying to break into prisons," she said. "If they were, security at prisons would be very different. So the way that you construct an idea of security . . . is very different if you're worried about people getting in and out."

The abject failure of Blizzard's established quarantines within *WoW* was due, in part, to these two unanticipated issues. You can't really fault them for it; standard epidemiological models don't account for them either. According to Nina, no one's ever looked at what happens when people get into and then leave a quarantine zone. "When you talk about people breaking into quarantines, that changes everything," she said. "And that was something we hadn't thought of."

Of course, when Nina talks about this, she frequently gets objections along the lines of "oh, they're only doing that because it's a game, and there's no risk. No one would ever do that in the real world."[6] She doesn't necessarily agree. Of course, no one she knows is naïve enough to believe that what people do in a virtual world maps *directly* to the real,[7] but again, human behavior is inherently unpredictable—and our history is rife with examples of poor judgment or sheer stupidity in the face of serious risk. Studying human conduct during a plague in a virtual world is a first step, a good initial approximation of something epidemiologists can't effectively investigate otherwise.[8] After all, they can't very well release a potentially deadly pathogen into a crowded metropolis and see what happens. Nina likes to use the analogy of mice models for drug trials. "There's no one in the world to whom you could say, 'well, we stuck that into a mouse and the mouse is fine, so it must be perfectly fine. Go ahead and eat it.'" It's a nice indication, but that doesn't mean it's safe.

There's another issue at work here, one perhaps more central to the debate—and completely missed by those who doubt the validity of extrapolating observations made in virtual worlds out to the physical. Nina explained. "I think the people who are objecting to the 'it's not the real world' aspect are failing to understand that gaming can be very emotionally involving, but completely separately from that, even if it was not, it would give us an actually well-fleshed-out and understood context in which to begin to ask these questions. I'm unconvinced by the 'well, it's not a perfect approximation therefore we shouldn't do it' argument."

It's a fair point, particularly if you bear in mind our discussion about the Proteus Effect in the previous chapter. Very briefly, the Proteus Effect illuminates the physical connection between gamer and avatar: When a player creates a digital representation of herself within an MMO or virtual world, she becomes emotionally involved with it, to the point where she can be physically affected by virtual events that impact her avatar. Creating and playing a character in *WoW* is a serious commitment of time and energy, and the longer you play, the more connected you get to that character—so death in the game, though not permanent, does have an impact, and most gamers don't risk harm to their characters lightly. This suggests that, as a result of that emotional bond, you *can* make some valid correlations between game-world and real-world behavior.

Case in point: over the course of Corrupted Blood's progression, Nina witnessed many instances of altruism and courage, in the form of healers putting themselves at risk while trying to cure others afflicted with the disease or lower-level characters trying to direct players away from dangerous areas. She also saw panic—people fleeing population centers and heading for remote, isolated sections of the world to wait out the plague. All these behaviors are common during real-world epidemics.

And there's something else she observed, which sheds light on a particularly dark facet of human imagination, one that correlates to the idea of "patient zero"[9] and the post 9/11 anthrax attacks: bioterrorism. Some players infected with the disease teleported into heavily populated areas to intentionally spread the disease to as many as possible. "People got really smart about figuring out how to cause the most damage to the largest number of people," said self-professed bioterrorist Robert Allen.[10] His *WoW* guild, domus fulminata, used classic terror tactics to create chaos and virtual death. Like a typical real-world sleeper cell, the small group blended into the general population, disappearing into the background and lying dormant until a weakness in the system gave them an opening to strike. Then, they moved in like suicide bombers, sacrificing themselves to inflict as many casualties as possible.

In an instance of life imitating art, Charles Blair, deputy director of the Center for Terrorism and Intelligence Studies, believes that *WoW* could provide a valuable window into the operation of terrorist cells in the real world. His group already uses computer models to study terrorist strategies and tactics, but *WoW*'s human players add a dimension of realism and unpredictability beyond the reach of even the best artificial intelligence systems. The game's main strength, according to Blair, is that it "involves 'real' people making real decisions in a world with some kind of [controllable] bounds," and he feels that studying players' actions in-game could provide valuable insight to intelligence analysts.[11] Whether the knowledge they glean will ever be good enough to affect counterterrorism policy is up for debate. However, it may get them asking different questions and could present them with variables they haven't previously considered. At the very least, video games and virtual worlds could become a useful tool in the struggle to understand how terrorist cells form and function.

GOOD ENOUGH FOR GOVERNMENT WORK: CDC GETS INTO THE GAME

When it comes to virtual epidemics, *World of Warcraft* isn't the only game in town. In 2002, three years before Corrupted Blood, a new highly contagious disease began spreading through the tween-/teen-centered virtual world *Whyville*. Founded in 1999 by Dr. James M. Bower and his students and colleagues at CalTech, *Whyville*'s primary purpose is to educate kids aged eight to sixteen on topics across a range of disciplines, including art, science, business, and geography. The illness—dubbed Whypox, for the chicken pox–like red welts that appeared all over afflicted Whyvillians' faces—was released intentionally by *Whyville*'s developer, Numedeon, to get the community engaged in learning about infectious disease.

Whypox only has two symptoms: the red welts already described, and a persistent, random sneeze that inhibits Whyvillians' ability to talk to one another. The way Numedeon chose to implement the sneeze was quite clever. Players communicate with each other by means of a text chat system: you type what you want to say into a chat window and others respond in kind. If you've caught Whypox, every so often "Achoo" appears randomly in the chat window, interrupting what you're typing, indicating to everyone that you're sick (in case they missed the welts), and prompting healthy Whyvillians to avoid you for fear of falling ill themselves. You also use chat to get around in *Whyville* by typing in the location you'd like to visit. For example, if you want to go to the mall, you'd type "teleport mall," and you'd be transported there instantly. If you've contracted Whypox, though, there's a pretty good chance that you'll sneeze at some point, so your chat might read "Achooteleport mall." As this is not a valid command, you won't end up traveling anywhere, and you'll have to retype it; if you sneeze again, you have to retype again. And so on.

No, it's not life-threatening, and no one's really getting sick. However, Whypox is far from trivial. Bear in mind that the two main activities kids use *Whyville* for are to design avatars and socialize with friends, and Whypox affected both—particularly the social aspect, as coming down with the virus could find you very quickly ostracized. Surveys conducted after a particular Whypox outbreak indicate that a majority of Whyvillians actually felt bad themselves when their avatars got sick, predominantly citing the impact to social interaction.[12]

There's no cure for Whypox; once you catch it, you just have to let it run its course. At most, you can visit the local *Whyville* pharmacy and purchase medicine to treat the symptoms. The only other option is to get vaccinated beforehand—and this is where the Centers for Disease Control and Prevention (CDC) got interested. A good portion of the CDC's mission is educational outreach, and in Whypox they saw an opportunity to, in a fun and nonthreatening way, instruct kids on the value of vaccination. They connected with Numedeon and agreed to partially fund further development of a *Whyville* epidemic, changing the name to Whyflu and launching the new bug in 2006.

For the first two seasons (2006–2007 and 2007–2008), Whyflu was not substantively different from the original Whypox, and your recourse to either prevent or treat the infection was the same (vaccination, isolation, treatment of symptoms). Starting with the 2008–2009 Whyflu outbreak, the CDC stationed one of their flu specialists in-world to answer questions and provide information about Whyflu and infectious disease in general, while Numedeon began incorporating preventative behaviors into *Whyville*: players could issue a hand-washing command to help keep their own avatars healthy—and vaccination, of course, was still an option. For 2009–2010, Numedeon modified the disease, adding a second strain—WhyMeFlu—with its own specific vaccine (mimicking the real-world appearance of H1N1, or swine flu) and another hygienic behavior that people could command their avatars to carry out. The vaccine for the original strain was also available, and washing your hands to stave off illness still worked. Now, though, if you contracted the disease, you could help keep it from spreading by covering your sneezes. This was the first time a sick player could take direct action to prevent passing Whyflu along to others, and it added another dimension to the interaction among avatars. As the CDC's Dan Baden told me during the eighth annual Games for Health conference in Boston, "there was social pressure on people to pick up these behaviors. So if you were not covering, people would chastise you." Not only did you run the risk of becoming a pariah if you caught Whyflu, now people would berate you for endangering others.

And it worked. Between the in-world education efforts led by the CDC and the social pressures placed on Whyvillians by each other, virtual vaccinations rose from about nine thousand in 2008-09 to more than twenty-seven thousand a year later; over that same time period, cases of Whyflu dropped from over seventy-seven thousand to less than fifty-five thousand.[13] The CDC also noticed extensive use of both preventative behaviors, suggesting that kids involved in *Whyville* understood both the benefits of getting vaccinated and of using hand-washing and sneeze covering to reduce the chances of becoming infected or transmitting the sickness to others. Whether this translates into real-world behavior change is unclear, but another of the CDC's recent efforts is showing some promising results. It's called virtual reality training, and it has the potential to dramatically affect the way rescue personnel prepare for everything from mineshaft collapses to natural disasters.

Traditionally, training for catastrophic event response is done at a specific location set up for a particular event—an earthquake, for example. It's run by a dedicated trainer, involves actors playing the roles of victims in various states of injury, and is very good for individuals or small teams. By all accounts, it's extremely effective: It offers a high degree of realism—second only to an actual disaster site—and gives trainees the opportunity to fail safely, as no actual lives are in danger. However, it's not easy to customize if you want to incorporate additional variables into the scenario (like a gas main explosion), is expensive to develop and run, requires a professional trainer onsite, and is very inefficient, both in terms of use of personnel—people have to be pulled out of other activities to stand in as victims—and in the small number of trainees who can run through it at any given time. Fortunately, technological developments over the last few years are now providing a much-needed alternative: immersive virtual environments.

Virtual reality environments (VREs) are essentially miniature virtual worlds that can contain anything you want to build into them: a neighborhood hit by a tornado, a train station destroyed by a suicide bomber, a caved-in mine, or even a simple two-car accident at a busy intersection. You can have as many or as few victims as you want, and they can run the gamut from barely scratched to near death. Like onsite training, VREs allow trainees to fail safely; however, they can also be customized fairly easily and can incorporate sight, sound, and even smell. The two biggest advantages of VREs, though? Flexibility and reach. Because you can deploy a VRE over the web, Baden told me,

> . . . you can have people doing the training at the same time or whatever time meets their needs. You can do distance training, as opposed to bringing everyone in to one location, having one instructor with six people or whatever their ratio is. . . . You have high instructor/trainee levels, you can vary the scenario much easier than having people try to imagine it, saying "okay, there's now a fire in front of you." I'm showing you a fire, and you're gonna have a more visceral reaction to this, and hopefully the lesson could be retained a lot better because of that visceral-type reaction.

That gut-level connection between trainee and scenario is part of the power of these environments. As we've seen on several occasions throughout our journey together, people respond to situations that happen to their avatars in MMOs or virtual worlds like *Second Life*. With VRE training there *is* no avatar between you and the action: everything that takes place is happening to you—it's just happening virtually. Images are more intense, events more realistic, and people react to them on a deeper level. It's like the difference between watching television and getting caught up in really good 3D: when something comes at you, you move.

The CDC tries to keep the scenarios pretty generic—earthquakes, hurricanes, dirty bombs, car accidents—running them as-is or adapting them to meet specific needs. They can also combine multiple scenarios—a car accident with an earthquake, for example—or develop one around a specific region—say a rural African village—to add a cultural dimension that may complicate matters and affect the CDC's response. "I know of one of the scenarios," Baden said, "I think it's a hurricane, you've got someone who's selling shrimp out of a cooler. And it's a good price, and hey, it's fresh shrimp, we're on the coast, it's gotta be good. Power's been out for four or five days, so this is a health risk. So it's got preparedness for the environment as well—what kind of situations they'll run into and may have to intervene in."

And then there's the classic event—one that rescue workers are all-too familiar with: poison gas trapped in a basement or well. One person goes in and doesn't come out, so a second goes in, and then a third . . . before long, you've lost half a dozen people, and what started out as a family crisis is rapidly becoming a community tragedy. You need well-trained rescue personnel who are adequately protected and prepared to deal with this scenario—both the physical threat of the toxic gas as well as the mental stress of panicked family members screaming at you that their loved ones went down there and someone needs to save them. VRE training can capture all of this, and it provides a way for the CDC to safely place trainees in dangerous situations, allowing them to experience and prepare for the confusion that attends an emergency.

Preparedness is only one part of the equation, though. To be effective during a crisis, you need people who can remain calm and function while surrounded by high levels of chaos, stress, anxiety, and fear. This is the province of what the CDC calls resiliency training—the goal of which is, essentially, to get people used to going into catastrophic situations and not freaking out. It's a type of training for which VREs are ideal. They're safe places for trainees to work in, allowing them to make mistakes or experiment with different strategies or approaches without putting lives at risk. And you can gradually amp up the intensity, increasing the tumult by adding more noise and distraction, heightening the level of panic, or complicating the situation with contingent factors they have to deal with.

Resiliency training operates on levels beyond just the individual, though. Yes, the primary concern is to ensure that rescue workers have the skills needed to effectively perform. However, they may be faced with situations where they have to support team members who haven't had the training. As Baden explained to me,

> When an emergency happens . . . say it's somewhere that they have some unique industry, and the person who's an expert in that industry hasn't been through the training. Well, they'll pull them anyway, 'cause we need that expertise. So it would be good to have someone who's versed in the resiliency training to help that person who's not prepared at all. And then the third level is these people can be more effective helping the victims of that emergency or disaster.

The question is, does it work? According to Baden, the answer is yes. The CDC has pre- and post-training information that clearly indicates this: Participants report that they're using their resiliency training in real-life experiences, and they feel more prepared when deploying to assist in a crisis. "They've said that they've used what they learned in the field," Baden said. The transference of skills is unquestionably real.

It's a phenomenon Jerry Heneghan's familiar with, albeit in slightly different context. Jerry's the founder of Virtual Heroes, a company that designs virtual reality training simulations for a wide range of applications (focused predominantly in the areas of health, public safety, and the military). They build scenarios that put trainees into the aftermath of catastrophic events—tornado strikes, terrorist attacks, combat operations, earthquakes, major accidents—and require them to manage the situation, identify victims, set up triage tents, check the scene for additional hazards, etc. Before starting Virtual Heroes in 2004, Jerry worked in a variety of game-related areas, and before that, he was on active duty in the US Army, serving as a company commander in a Korean demilitarized zone, where he also flew Apache helicopters. He was in a fairly remote location, and during downtime, he'd play video games—especially flight simulators—on a PC his brother had sent him. "I became fanatical about playing games," he told me, "and I was fascinated that the games that I could play on my earlier Pentium computer were better, in terms of fidelity, than the simulator that I used to have to use in the Army that cost about a thousand dollars an hour to run."

Flying an Apache requires very precise movements, takes a long time to get used to, and is a skill that demands regular practice (it falls firmly in the camp of "use it or lose it"). For a variety of reasons, there were times that Jerry was grounded for a week or two.

> Normally when you do that, you're pretty rough the first night you go out there. But I got hooked playing these games, and I played all the Apache games, and I would go out and the guys that I would fly with would be like, "God damn, you're in the groove. My God, you haven't flown in what, a week or something? Holy cow!" But I logged, like, fifty-six hours playing the Comanche game.[14] And I played the whole game and started to see a correlation there.

That correlation between the two—the transfer of skills from simulation to reality—combined with Jerry's lifelong love of games informed his professional decisions post-military, eventually inspiring him to launch Virtual Heroes. One of the first projects that the nascent company worked on was an update of the first-person shooter *America's Army*. Designed primarily as a recruitment tool (more on that later), the game gives players a sense of what life as a soldier in the US Army is like. As part of the game, players can choose to take on the role of combat medic. However, you can't just step into the job. You first have to pass a rigorous virtual medical training course that's based on actual

US Army medic training, covering everything from evaluating casualties and controlling bleeding to treating shock and resuscitating nonbreathing victims—and it's this aspect of *America's Army* that gave Jerry his first taste of what a video game could do.

One Saturday, Jerry got a call: Virtual Heroes was on CNN. His first thought was, "what did we do now?" So he tuned in to listen.

> Some young kid in North Carolina, on the highway—I-95—car rolls into a ditch in front of him, minivan, family, and they're thrown all over the place and messed up—broken bones and bleeding—and he, seventeen years old, jumps out, calls for help, starts identifying the victims, makes sure he's got everybody accounted for, applying pressure, dressing, treating for shock, calming people down. And because of where they were, it took fifteen or twenty minutes for someone to get there, and they realized after the fact, well, if this kid hadn't been there, these people probably would have bled to death.

How did he acquire those skills? By going through the combat medic training in *America's Army*. It's not the only case of someone applying skills learned in the game to a real-world situation: another North Carolina man saved two people when their SUV went off the road in front of him.[15]

That kind of skills transference is exactly what Jerry hopes to achieve with HumanSim, one of Virtual Heroes' flagship health care simulations. Launched in late 2011, HumanSim provides health care professionals a realistic, immersive virtual environment in which to learn, practice, and refresh their skills without putting patients in harm's way. Virtual Heroes used cutting-edge game design systems to create the highest degree of visual fidelity and realism possible, incorporating a highly complex and accurate physiology engine to drive the virtual humans that populate each training scenario, allowing for incredible variability in symptoms and conditions. Biomedical experts on staff and medical professionals from the Duke University School of Medicine teamed up to provide HumanSim's programmers with real-world data based on their own experiences and research, ensuring that patients react realistically to the conditions around them: Virtual victims behave and degenerate exactly the way that real victims would—for example, when injured, they lose blood and use oxygen at the same rate.[16] The result is a dedicated virtual training environment that promises to drastically and permanently alter the health care landscape.

Of course, there's more than one way to start a revolution. While Jerry Heneghan and Virtual Heroes are driving the advancement of medical simulations, laparoscopic surgeons have discovered that sometimes affecting change is simply a matter of knowing what game to play.

Laparoscopic surgery involves making tiny incisions in the body and inserting a remote-controlled video camera that transmits images to an external screen. Procedures are minimally invasive and accomplished by the use of miniature surgical instruments. By watching the video screens, surgeons navigate cameras and instruments through the body using video game–style controllers. Joint studies conducted by Dr. James Rosser and Dr. Douglas Gentile of Beth Israel Medical Center in New York and Iowa State University, respectively, compared the performance of gamer and nongamer surgeons, and came up with some surprising results. Even after accounting for differences in age, years of training, and surgical experience, they found that gamers executed laparoscopic procedures significantly better: They made more than a third fewer mistakes and successfully completed advanced surgeries 27 percent faster than their nongaming colleagues.[17] According to Dr. Gentile, "the single best predictor of their skills is how much they had played video games in the past and how much they played now. Those were better predictors of surgical skills than years of training and number of surgeries performed."[18]

An Italian study went one step further, running an actual trial as opposed to a study of game-playing habits. Researchers at the Sapienza University of Rome compared resident surgeons who were asked to play one hour of Nintendo's Wii each day for a month against a nongaming control group; both groups had minimal experience with both laparoscopic surgery and the Wii. The results, according to research team leader Dr. Gregorio Patrizi, were astounding: at the end of the month, residents performed a simulated laparoscopic procedure, and those that had played the Wii fared far better. "The differences in outcomes between the two groups were far beyond our expectations," he said. "What surprised us the most was that almost all the results were clearly statistically significant, even in complex procedures."[19]

Most surprising, though, are results from the University of Texas Medical Branch at Galveston. After watching his ten-year-old son (yes, a gamer) master a robotic surgery simulator—having had no prior experience—Associate Professor Dr. Sam Kilic decided to see whether playing video games could help surgeons develop their robotic surgery skills. To make things interesting, Kilic worked with three distinct groups: high school sophomores who played two hours of video games a day, college students who played four, and experienced resident physicians who didn't play at all.[20]

So what happened? Both the high school and college students matched, and in some cases exceeded, the skills of the resident physicians. This caught Kilic quite by surprise, as his residents had already been actively operating. "I was thinking," he said, "[that] their surgical background would surpass the non–medical-field people."[21] On the face of it, that's exactly what you would expect. It turns out, though, that experience with the controller was more instructive than time on task. In this case, it wasn't that the game had taught the students the intricacies of robotic surgery. Rather, their familiarity with video game controllers allowed them to adapt quickly to the interface with the simulator.

Games can teach, though. And while it my take some time for them to become universally embraced as an educational tool, teachers are finding ways to incorporate games into the classroom experience.

Back in 2003, sixth-grade science teacher Cathleen Galas[22] and a colleague of hers at the University of California, Los Angeles, Dr. Yasmin Kafai, conspired to infect her students with a nonfatal but highly contagious pathogen. In most cases, this would be extremely unethical, and even criminal, but if you've kept up with me this far, then you've already guessed that this happened entirely within a game world—*Whyville*, to be precise. Years before the CDC began spreading Whyflu throughout the virtual populace, Galas realized that the social aspect of *Whyville* provided the perfect environment for her students to learn about infectious disease. She and Kafai contacted Numedeon (*Whyville*'s owner), and they agreed to schedule a Whypox outbreak to coincide with a National Science Foundation study on collaborative, immersive virtual environments. And so, in the fall of 2003, Cathleen Galas' class began to fall frustratingly, annoyingly ill.

At first, her students reacted to the social impact—appearance and conversation: The red welts that attended the illness made their avatars look terrible, and the random, constant sneezing bothered them because it made chatting difficult. Those not yet infected (or who had recovered) worried about catching the disease or becoming re-infected, but also began to think about ways they could help their sick friends. Then the students started discussing how to control the spread of the infection—and that's when the exercise got really interesting. They shared data among classmates and also began collecting it from the larger *Whyville* community (it wasn't just Galas' class that was coming down with Whypox), tracking new infections as well as the progress of existing infections and looking for the root cause of the outbreak. She and her students created maps of the disease, charting its progress and continually adding information as they learned more about the spread of Whypox (and infectious disease in general). Galas provided supplemental information, and students took it upon themselves to research topics from a variety of sources to gain a better understanding of epidemics—becoming, in essence, "epidemiological detectives."[23] Tools within *Whyville* (at the in-world CDC office[24]) allowed students to model and run epidemic simulations, and watch how a disease spread throughout a community. As they learned more about infectious outbreaks, students evolved from detectives to activists. Galas wrote,

> They took to heart attempting to solve the real-world questions about Whypox and its spread through the *Whyville* community. Students organized online groups dedicated to research, education, and philanthropy. They actively spread the word about ways to prevent infection, education, organizing a hospital and university as future educational and research centers. They took on different tasks to combat the spread of the disease within their online community.[25]

The beauty of *Whyville* and the Whypox epidemic was that students were learning science by active investigation, research, and collaboration. They were responsible, in large part, for directing their own learning, which got them excited and engaged in the process in a way beyond the scope of any passively consumed classroom lecture or subject-matter video (no matter how educational). "This nontraditional science-learning environment provided a meaningful context for studying about epidemiology and problem solving," Galas wrote. "Students worked collaboratively beyond the borders of their classroom to investigate and solve the Whypox dilemma."[26]

Video games can do more than just engage students in learning, though. At their best they can address global challenges and help us to understand people and issues outside of our own narrow frames of reference and experience.

Game the News does exactly that. Not a game itself, Game the News (GTN) is a web portal for games that tackle a variety of current and/or newsworthy affairs. A project of Auroch Digital Ltd., the creative team behind GTN describe themselves as "the world's first news correspondents who cover global events as games. As news breaks, we create our own twist on events in a playable form."[27] These games fall solidly into the casual category—easy to play, hard to master—requiring only a modest time commitment and allowing designers to get them out quickly (development typically runs from a few days to a couple of weeks). They run the gamut from irreverent, as in *Cow Crusher*—a commentary on first-world food production developed in response to the horse-meat scandals of early 2013—to thought provoking, like *Climate Defense*, which asks players to balance a variety of factors to keep climate change in check, or GTN's current best offering, *Endgame: Syria*, which explores Syria's ongoing civil war. In *Endgame*, you play the part of the rebels, and your objective is to bring an end to the conflict by weakening support for the regime to the point where it offers a cease-fire favorable to your side. To succeed, you must use a combination of military and nonmilitary options (sanctions against the regime, armed resistance, political pressure) while balancing public opinion, civilian casualties, global support (for both the regime and the rebels), and your own limited resources. Your choices at each stage affect your options later in the game and ultimately the entire outcome: Incur too many civilian casualties, and you'll lose support; block the regime from acting against you, and support for your side goes up. It's easy to play, but difficult to win, and it does a good job of getting you to understand the complexity of the situation and the challenges involved in resolving it—certainly better than simply reading about it or watching it on the news.[28]

Game the News isn't the only group creating games around issues like these. The concept of games for social change is now well established, and those trying to employ games to raise awareness and generate positive attention can find developers ready to build them and audiences willing to play. Two of the most notable efforts (as of 2013, at least) are *Darfur is Dying* and *Half the Sky*.

Led by University of Southern California grad student Susana Ruiz and launched by mtvU in partnership with the Reebok Human Rights Foundation and the nonprofit International Crisis Group, *Darfur is Dying* provides a window—however briefly—into the experience of the nearly 2.5 million people directly affected by the bloodshed in Sudan.[29] You play the game by navigating a Darfurian refugee (you have a choice of eight: two adults and six children) through a variety of situations that threaten the survival of both you and your refugee camp. Along the way, you have the opportunity to learn about the history and current status of the crisis and genocide, as well as discover how you can help.

Half the Sky grew out of the groundbreaking work on women's rights by journalists Nicholas Kristof and Sheryl WuDunn. It began with their 2009 best-selling book *Half the Sky: Turning Oppression into Opportunity for Women Worldwide*, developed into a television series in late 2012, and turned into a movement aimed at nothing less than ending the global oppression of women. As a part of this effort, Games for Change (one of the leaders in the creation and distribution of social impact games) is overseeing the development of three mobile games, aimed at reaching the 3.5 billion mobile phone users on the planet. In order to leverage the power of social media, the *Half the Sky* team (again, in partnership with Games for Change) also released a Facebook game early in 2013. You play the role of the protagonist, a woman named Radhika (based on a real Indian woman), who you lead on a journey across the globe, facing challenges, interacting with a variety of characters, and completing a series of quests—each one based on actual work done by one of *Half the Sky*'s seven nonprofit partners.[30] At the end of each quest, you'll be invited to get involved with the nonprofit at the center of the issue. Throughout the game, you'll also have the chance to complete tasks and unlock various actions; in a first for Facebook, those actions have direct real-world effects: collect books for girls in the game, and you'll activate a donation of books to the real-world organization Room to Read. Players can also make direct, personal donations to any of *Half the Sky*'s partners at any point in the game. Of course, it's still just a game, and no one's sure how much impact it will really have. Reality is harsher than the game world presents, and the *Half the Sky* team knows it. They hold out hope, though, that it can at least make people aware of the issue, and by doing so, start to make a difference.

All of these games elevate the idea of social activism from a passive, one-way exchange to an interactive conversation between player and event—providing opportunities to educate ourselves and act to drive change. Through the experiences we have, we gain insight into these events and perhaps grow to care about those affected by them.

But if games can get us to care, can they teach us to kill? It's the perennial question: do violent video games turn us—and particularly our youth—violent?

Before I take on that question, allow me to interject a little personal experience here. As I mentioned previously, I grew up playing video games—at arcades, at friends' houses, at home—and so did most of my friends. And I played *a lot*. Still do, actually, though not nearly as much as I used to—the realities of adulthood have drastically reduced my gaming time. In recent years, I've also met quite a few people—adults, kids, men, and women—who game, some seriously. Many play first-person shooters, myself included. Most of us are successful and well adjusted. And as far as I know, none of us have become raging sociopaths as a result. If video games could fashion us into cold-blooded killers, you'd think that one of us would've snapped by now. Extend that out to society at large, and the chances for game-driven tragedy should go through the roof: According to the Entertainment Software Association, the average US household has at least one computer or game system used to play video games.[31] That's a lot of potential killers. And yet here we are.

Still, video games are a common scapegoat, trotted out as one of the great evils of society whenever some otherwise inexplicable act of violence makes the news. This is nothing new. Emergent media has always suffered from suspicion and accusation—comic books, radio, TV, and movies have all been viewed by many as harbingers of doom, heralding society's imminent collapse. Video games are just the latest target, the next medium in line to serve as the proverbial hand basket in which we're all going to hell.

The evidence does not bear this out. If you look at US Justice Department crime statistics, violent crime per thousand people—a measure that accounts for population growth—has dropped precipitously: between 1993 and 2010, it fell 70 percent.[32] At the same time, video games were exploding in popularity. It's also worth noting that the decrease began in 1993, the same year that both *Doom* and *Mortal Kombat*—two of the most popular and vilified violent video games in history—were released.

Those are overall crime statistics, though. They're just numbers, and they don't say anything about any specific activity. And they certainly don't get us into the minds of gamers. What about scientific studies of people playing these games? What do the experts say about the effects of video game violence and the link between violent games and violent behavior?

The short answer is: there is none. In the nearly four decades since Exidy's *Death Race* touched off the first violent video game controversy in 1976, not one study has been able to prove a causal relationship between violent video games and real-world violence.[33] And it's not for lack of trying: Over the years, several studies by high-profile researchers (most notably Iowa State University's Dr. Craig A. Anderson) have claimed to make this link. Invariably, though, those studies don't hold up to scrutiny. According to Christopher J. Ferguson, Associate Professor of Psychology and Criminal Justice at Texas A&M International University,

> . . . much of the literature on video game violence has relied on measures that have not been clinically validated as indicators of real-world aggression, are often created ad-hoc, and lack standardization. This continues despite the fact that it has been known for some time that better validated measures of aggression that more closely measure the construct of interest tend to produce much weaker effects in media violence research. This finding has been confirmed by analyses of video game violence research wherein poorly validated and, particularly, non-standardized measures of aggression tend to produce much higher effects than do clinically validated standardized measures of aggression.[34]

In their 2008 book, *Grand Theft Childhood*, Drs. Lawrence Kutner and Cheryl Olson—co-founders (along with Dr. Eugene V. Beresin) of the Harvard Medical School Center for Mental Health and Media—put it more bluntly. "The strong link between video game violence and real world violence," they wrote, "and the conclusion that video games lead to social isolation and poor interpersonal skills, are drawn from bad or irrelevant research, muddleheaded thinking and unfounded, simplistic news reports."[35]

Part of the problem is in defining exactly what we mean by violence and identifying markers that may be indicative of violent tendencies. I spoke with Massachusetts General Hospital's Dr. Atilla Ceranoglu about this and about where the state of research was today. Most of it, he said, was flawed.

> It uses vague definitions, like violence, violent behavior, aggression, antisocial behavior—these are not all exactly the same, or don't have the same cognitive or emotional processes. Nonetheless, it's all clustered together and studied in that context. . . . They look at irrelevant parameters or markers. That is, there are physiological markers, they look at cognitive markers, they look at . . . these markers are for violent behavior or for violence or for act of aggression or aggressive state. And where . . . the physiological marker they propose is heart rate increases. Well, heart rate increases, true, during aggression or violence, but heart rate increases at other times, too.

What does concern him is a child (or adolescent) who plays violent video games to excess and does little else—doesn't go out and socialize with friends, isn't playing other games. In this instance, Dr. Ceranoglu believes that the attraction to violent video games might be a marker for someone who's at risk of actually becoming violent—or who may already be. It's a thought echoed by Chris Ferguson: His research has shown evidence that children who are already hostile or aggressive may become more so by playing violent video games.[36]

Villanova University psychologist Patrick Markey's research also supports this idea. He found that children whose personalities showed high degrees of neuroticism and disagreeability became slightly more hostile through playing violent video games. For kids who are moody, impulsive, or unfriendly, allowing them to play such games is probably not the best choice a parent could make. "Video games are not simply good or bad for everybody," he said. "But for some individuals who have certain dispositions, if they play video games they're much more likely to be negatively affected."[37] That being said, such kids probably shouldn't be exposed to violent media of any kind—movies, TV, even the news. Regardless, Ceranoglu, Ferguson, and Markey are of one mind on this point: there is no hard evidence pointing to violent video games as a general source of violence or aggression.[38]

So, if the link between violent games and actual violence is so tenuous, why is it continually cited? Why, when we're faced with horrific events like Columbine, Virginia Tech and, most recently, the Sandy Hook Elementary School shooting, do we fall back on the "games are making us violent" mantra?

Because it's easy. Because we want answers. Because we don't know what makes someone violent, and that scares us.

I wish it were that easy. It would be wonderful if we could find such a simple reason. If we could identify a cause, we could stop it, and take comfort knowing that we've eliminated the one thing that can turn our kids from kind, loving children into psychopathic purveyors of death.

Sadly, it's not. Sandy Hook reminded us of that. A troubled youth vents . . . what? Rage? Frustration? Fear? Hopelessness? We'll never know. Whatever the reason, it left twenty-six dead, mostly children who, just minutes before, were full of life and joyful anticipation, as only the youngest of us are. For days afterwards, I wandered around in a haze, unsteady on my feet, seeking something solid to anchor myself to, taking comfort in my own family while lamenting those whose families had been torn apart. In the wake of the tragedy, we were all casting about, looking for some way to explain the senselessness. Looking for an answer.

I wish I had one to give. I do know this, though: we'll never find one unless we start asking different questions. When we look to place blame on any external factor, be it violent video games or something else, we distract ourselves from getting at the root of the issue, and facing some truly difficult questions that force us to examine what we, as a society, value and what lessons we teach our children as a result.

Suffolk University professor Nina Huntemann and I talked about this one afternoon a few years earlier. "I think about all these cultural artifacts," she said, "movies, video games—I think about them as all part of a learning environment. We're just constantly reading the messages in all of these artifacts and adding them to our understanding about the world. And it is unfortunate that we live in a world right now that glorifies violence."

In this context, she said, "video games are just another medium that participates in the conversation, another medium that replicates the glorification of violence." It's that conversation—about what we, as a culture, believe in and hold most dear—that we must have. And it's not easy. "That makes parents responsible," Nina said, "it makes the producers of video games responsible, it makes politicians responsible, educators, doctors, everybody is responsible for the culture that we decide to participate in. And that's a harder thing to change than simply banning a video game from a shelf."

In the final analysis, perhaps what most upsets people about video game violence is not that it drives human behavior, but that, like a mirror, it reflects those aspects of human nature of which we're least proud. Right now, we live in a society that accepts violence as the norm, even celebrates it. There's a certain level of comfort with it; you see it echoed in nightly news reports (how many were killed?), in movie ratings where sex is more regulated than violence, and in national policy, when we restrict access to health care and education, deny equal protection to loving same-sex couples, and refuse women the right and freedom to control their own bodies, yet cry foul at any attempt to even discuss sensible regulation of firearms. We would do well to consider the message this sends.

When faced with a difficult problem, it's comforting to point to an external indicator and say, "yes, that's it. That's the cause." And it's facile to think that removing it will somehow create a lasting solution. But this only treats the symptoms, it does nothing to cure the disease—and unless we can do that, we will continue to fall ill, with tragic consequences.

It also places responsibility comfortably on some outside force, allowing us to turn the mirror outward, away from ourselves. We can breathe easier, secure in the knowledge that it's not us. By doing so, we avoid asking deeper questions whose answers, though potentially frightening, might let us finally understand who we are and what we must do to change. Violent video games are not the source of our ills, and censoring them won't resolve the underlying issues, whatever they may be. It will only serve to cloak the mirror in darkness again, allowing us to ignore it until tragedy inevitably lifts the veil and forces us to once more confront the quality of our society, our culture, and our humanity.

10

GAMES FOR HEALTH

The year is 2027. Just three years after the greatest medical advance in history, scientists are poised at the edge of yet another breakthrough—and this one looks to be even more powerful. By 2024, nanotechnology had arisen as an effective means of gene therapy and drug induction. Combining it with the results of decades of research in another groundbreaking discipline—artificial intelligence—scientists achieved something so profound that it would fundamentally alter the practice and delivery of health care forever: self-aware, artificially intelligent nanobots. Fully autonomous and able to operate from preprogrammed data or learn on the fly, these microscopic robots were quickly adapted to fight disease and could do so with a degree of precision and control previously unimaginable. They attacked disease at its source: the cells themselves.

Scientists tested the first generation of self-aware medical nanobots—the SMT-100—in 2025 as part of the newly formed nanotech chronic illness treatment program. Targeted at patients with cancer, this series met with moderate success in human trials. Encouraged by early results but still not satisfied, they went back to the lab and spent two years reworking and redesigning. The result of their efforts was Roxxi—the RX5-E series medical nanobot, a piece of technological wizardry unmatched in human history, designed for a single purpose: seek out pathogens and destroy them. Like the SMT-100 before her (though mechanical, Roxxi's appearance and bearing were distinctly female, and she identified as such), Roxxi was directed against cancer. The question was, would she work?

It was time to put Roxxi through her paces.

From the results of her first trial, scientists had their answer. Injected into a test group of cancer patients, the nanobot's performance was astounding: No larger than her targets, yet armed with a brace of radioactive and chemical weapons, Roxxi tracked and annihilated malignant cells with startling efficiency, leaving healthy ones intact and unharmed.

After more than a century of dedicated research, heartbreak, and toil in the fight against cancer, scientists had finally found their weapon.

• • •

For most of us, the concept of intelligent, microscopic robots fighting disease inside the human body is an idea confined to the realm of science fiction. However, there is a small segment of the population that's experienced this firsthand—not by being injected with nanobots, of course, but through the magic of a video game called *Re-Mission*. The game was designed for a specific audience and a set purpose: to get kids with cancer to understand and follow their treatment protocols. Why magic? Because it works, and for these kids, that makes all the difference.

Re-Mission isn't alone. It's part of a growing movement around developing video games that do more than just entertain us, but that make us happier, make us healthier, and may even save our lives. They're called serious games, and one of their most visible and diverse applications is in the area of games for health.

Serious games sounds like an oxymoron, a contradiction in terms. How can something inherently designed for play be serious? Yes, there are serious *gamers*, but the games themselves are only for entertainment, for killing time. Even those that foster social interaction are still for fun. And games for health? That sounds even more suspect. If you've stuck with me this long, you might be willing to grant me that video games have some positive aspects, particularly around social connection, education, and training. But the idea of a video game actually making you healthier may still be hard to swallow. When it comes to physical or mental well-being, they can't really help, right?

Wrong.

It turns out that health games can provide a variety of physical and mental benefits, and even drive behavior change in a life-saving direction. Let's not get ahead of ourselves, though. Just as it's important to begin a strenuous workout with a good warm up, let's ease into our discussion of health games by starting with something easy to grasp: exergaming.

Exergaming is exactly what it sounds like: exercise through gaming. If you're at all familiar with the Wii, Xbox Kinect, or PlayStation Move, then you at least know a bit about this. Both the Wii and Move controllers respond to motion (through either the Wiimote's internal accelerometer, or the Move's combination of handheld remote and motion-capture system); the Kinect is essentially a motion-capture camera that picks up full body movement, allowing gamers to ditch the thumbsticks and instead use their bodies as game controllers. All three of these systems—and, it seems likely, their descendents well into the future—encourage gamers to get up and move. Titles like *Wii Sports*, *Wii Fit Plus*, *Kinect Sports*, and *EA Sports Active 2* were designed around the idea of making real exercise enjoyable and nonthreatening—what you might think of as dedicated exergames. Others, like *Dance Central*, *Just Dance*, and *Dance Dance Revolution* (*DDR*) are games first, exercise programs second. They provide what I call unintentional exercise: playing these games requires you to move, but they're fun and engaging, and the movement is an intrinsic part of the gameplay, so you don't realize that you're actually working your body.

But how much are you doing, really? Like anything else, what you get out of it largely depends on what you put in. From my own experience, I can tell you that exergames work. In early 2012, Electronic Arts sent me an evaluation copy of *Sports Active 2* (*EASA*)—at the time, their most advanced exergame—which I dug into with great enthusiasm. I've spent the better part of a year working with it, and I can tell you this: it may be a game by classification, but it's serious exercise by any other measure. Designed by a team of game developers and certified professional trainers, *EASA 2* includes more than seventy different exercises and fitness activities that target all the major muscle groups—core and upper and lower body—and cover everything from flexibility and strength training to cardio and aerobic health. It comes with several preprogrammed, canned workouts that, on their own, will very effectively take you from novice to fitness expert. *EASA*'s real power, though, lies in its flexibility. You can select from its full menu of exercises and build a workout to meet your individual needs, or create several to cycle through, perhaps targeting a different area each day of the week. If you prefer, *EASA* also has the ability to generate workouts on the fly, based on the amount of time you have and what you want to focus on. And you can vary the intensity of either the entire workout or specific exercises, so you can gradually and safely develop areas of weakness and push yourself where you're stronger.

By using nothing more than *EASA 2*, I've strengthened my upper and lower body, increased my aerobic capacity, and even rehabbed a muscle injury to my arm. Without a doubt, the program provides a real workout[1]—as real as any of the glut of exercise videos on the market. It goes far beyond what any of them can offer, though, affording a degree of variety and customization beyond even the best video's wildest aspirations. And the system tracks everything: heart rate, calories burned, miles traveled, reps, number and duration of workouts, and lifestyle and nutrition info (through surveys you can fill out). There's even a virtual personal trainer (a choice of two, actually) to guide you through the exercises and help keep you motivated. Perhaps most importantly, you appear in the program as well, in the form of an avatar you create as part of your personal profile. This is powerful: Not only do you see your avatar performing the exercises, you get immediate visual feedback as to how well you're doing. When I'm exercising, I push myself harder to ensure that my digital self succeeds—and I get a better workout as a result.

Since their first serious appearance in the early 2000s, active video games (as exergames are sometimes known) have taken off. There's quite a bit of excitement and energy around exergaming now, with studies demonstrating that games like *Wii Sports* (particularly boxing and tennis) or *DDR*[2] provide as much or more of a workout as moderate-intensity walking.[3] Dr. Bruce Bailey at Brigham Young University and Dr. Kyle McInnis from the University of Massachusetts found that kids who played exergames for ten minutes got a workout as good as or significantly better than a ten-minute walk at three miles per hour on a treadmill. In the March 7 issue of *Archives of Pediatrics and Adolescent Medicine*, Bailey and McInnis wrote that "exergaming has the potential to increase physical activity and have a favorable influence on energy balance, and may be a viable alternative to traditional fitness activities."[4] Of course, they're both quick to point out that it shouldn't replace those activities, just enhance them. "Although exergaming is most likely not the solution to the epidemic of reduced physical activity in children," they noted, "it appears to be a potentially innovative strategy that can be used to reduce sedentary time, increase adherence to exercise programs, and promote enjoyment of physical activity."[5]

It's precisely that enjoyment, the entertainment appeal of these games, that makes them so effective. Kids really have fun with them and are more likely to stick with the program—and reap the benefits—as a result. George Velarde agrees. He's the chair of the phys ed department at Sierra Vista Junior High in Canyon Country, California. In 2003, he added an exergaming room to the school's fitness center. "The kids don't even know they're working out," he said, "but they are working out even more at moderate to vigorous levels because of exergaming."[6]

Which is very, very good, because the US population is growing, and I don't mean our numbers. There's no delicate way to say this, but too many of us are fat—really fat. Obesity in this country is an epidemic: according to the Centers for Disease Control and Prevention, about a third of all adults and 17 percent of children—three times the rate of twenty years ago—are obese,[7] and as of 2013, not a single state in the union has met the Healthy People 2010 goal to lower obesity rates below 15 percent.[8]

Being obese increases your risk of heart disease, type 2 diabetes, cancer . . . the list goes on. And its economic cost is staggering: Obesity hammers us with an annual medical bill of around $140 *billion*, and it's rising:[9] by some estimates, additional health care spending on obese Americans could reach more than half a *trillion* dollars over the next twenty years.[10] I don't know about you, but I can't even conceive of that amount. Most distressing of all, though, is this: due to the precipitous rise in childhood obesity, for the first time in US history, today's generation of kids may not outlive their parents.[11] It's absolutely critical that we reverse this trend—our physical and economic survival depends on it. Getting us off the couch and moving is a good place to start, and active video games are an integral part of the solution.

That video games are even included as part of the conversation around combating childhood obesity is an encouraging sign and a sea change from where we were just six years ago. In 2007, video games were still considered part of the nation's weight problem and exergaming as a word didn't even exist. New Mexico State University's Dr. Barbara Chamberlin remembers it well. "I submitted a grant proposal suggesting we study the potential of video games to increase physical activity, provide motivation for exercise, and take advantage of social, health-related activity," she said. "My proposal was not well-received."[12] In fact, she was taken to task for submitting it in the first place and highly encouraged to never do something so foolish again.

Today, the three major gaming consoles (Xbox, PlayStation, and Wii) all have exergaming components, and more and more people are looking into the physiological and social effects of video games.

And then there are those bold, intrepid souls who've taken video games out of their somewhat comfortable domain of exercise and weight loss and brought them into far murkier realms. Like digital Magellans, they're seeking out new routes to healing the body and the mind, and they're adapting video games to do it.

FROM REHAB TO WII-HAB: USING VIDEO GAMES TO HEAL

My great aunt Rose is pushing eighty-five. She's lived in Brooklyn all her life. She has Parkinson's disease. And she'll totally own you on *DDR*.

Okay, not really. But after a few good sessions of *DDR*, the outward signs of her disease begin to temporarily vanish. And she's not alone: Rose is one of several subjects in one of many studies looking into the ability of video games to repair the mind.

It was indirectly through Rose that I met Dr. Adam Noah. At the time, he was teaching neuroscience at Long Island University's Brooklyn campus and heading up the school's ADAM Center,[13] a teaching lab that explores the biology, physiology, and mechanics of human movement. Adam is a neurobiologist and lifelong gamer, and these two abiding passions dovetailed into his current research on the ability of video games to manage disorders of the brain.

Adam hit upon the idea of treatment through an active game—specifically *DDR*—after working with patients who'd suffered damage to their spinal cords. The gold standard for rehabilitating spinal cord injuries or motor-based neurological disorders like Parkinson's is treadmill training: get them exercising on a treadmill, rebuild their strength, and reduce their risk of falling. Historically, it's been pretty effective. The problem is, people plateau—they reach a point where treadmill training's gotten them as far as they can go. And the question is, what happens next?

Often, the patient moves on to training real-world tasks, like walking on carpet or snow. But Adam, who's a *DDR* aficionado, suspected that active video games might be a possible extension of the traditional rehabilitation model. "I always thought that it might make more sense because most of the people who get spinal cord injuries . . . tend to be young males," he told me. "And so what do young males like to do? They like to play games. So I'm thinking, 'well that treadmill therapy looks awfully boring, and I don't like running on a treadmill myself. What do I like to do?'"

Adam was living in Canada at the time and had a *DDR* unit in his apartment. He'd encountered the game on a trip to Japan and fell in love with it, so he bought the home version and started playing around with it. "It'd get to minus fifty in Canada, on a regular basis," he said, "so I needed something to do inside besides running my treadmill. So I started playing the game for exercise, and then I started to put two and two together . . . you know, I'm getting some skill here that I didn't have before, and I wondered if that could be applied to an injured population."

Shortly after that realization, Adam was struck with another, more pressing reality: the expiration of his Canadian visa. He returned to the United States and Long Island University, bringing his *DDR* and his interest with him—and that's when he met Rose, who became his first pilot subject (her husband Harvey was a somewhat more reluctant subject number two. Others followed).

The goal for treating any neurological disorder is to reduce the risk of falling. As Adam explained, "One of the major things is falling down can cause a cascade of events, which really ultimately can lead to death. That's the main thing that people that have Parkinson's die from, is actually falling down."

Typically, the progression goes something like this: a Parkinson's sufferer falls down, breaks his hip, and ends up in the hospital. Complications add up and he's not healing well, so he winds up in a home, where, over time, he slowly deteriorates until the end. Reduce that risk, and you can prolong someone's life. "And I always thought that playing this silly little game—*Dance Dance Revolution*—could just reduce the risk of falling. And that's all it was about."

There are a number of factors associated with that risk like strength, confidence, mental awareness, cardiovascular fitness, and reaction time. Many traditional intervention methods address one or two of these; when Adam looked at what *DDR* was actually doing, he realized that the game targeted most, if not all, of them. And when he and his team reviewed videos and began analyzing the data they were collecting, they saw improvement across the board, in all risk factors. They'll need to conduct long-term studies to determine the implications of these results, but Adam is confident. "If you put two and two together," he told me, "if you reduce the risk factors, you should reduce the overall risk of falling, which should reduce the incidence of falls." If he's right, he may have found a key to help those with Parkinson's manage their disease, stay out of the hospital, and ultimately live longer, healthier lives.

Parkinson's sufferers aren't the only ones for whom active video games have become a boon. Game technology is transforming traditional physical therapy from a necessary evil to an endurable, and even enjoyable, process.

The experience of physical therapy—or "pain and torture," as it's affectionately known by many who've endured it—can range from tedious and uncomfortable to excruciatingly painful and terrifying. The catch is it's essential to recovery. So how can you take a treatment modality that's vital, though often unpleasant, and make it bearable?

Easy. Occupy the patient's mind by providing it an entertaining distraction. In other words, turn it into a game. Dr. Gustavo Saposnik, a neurologist and director of the Stroke Outcomes Research Unit at St. Michael's Hospital in Toronto, has been studying clinics and hospitals that do just that—incorporating Wii gaming into the rehab regimen of stroke patients. Citing the results of one particular study comparing patients who used the Wii versus those who did not, Dr. Saposnik said, "we found that patients in the Wii group achieved a better motor function, both fine and gross, manifested by improvement in speed and grip strength."[14]

The speed gain was, in Saposnik's opinion, significant: They consistently edged their fellow nongaming patients by seven seconds when performing various tasks. "In other words," Saposnik said, "imagine that you have for every task you are doing, instead of doing that in twenty seconds, it will take you twenty-seven seconds for each activity. . . . That would be an impressive prolonged time."[15]

He suspects that the Wii's effectiveness lies in its mirroring of traditional rehab vis-à-vis carrying out repetitive, high-intensity tasks—the difference being that the Wii makes it fun, and that seems to be the key. That the Wiimote senses and responds to motion also gives it a higher degree of interactivity than other game consoles (the Xbox or PlayStation, for example), and also seems to make a difference.

There's another element at work here, and it's something for which, in this context, the Wii is ideal: empowerment. A few years ago, a friend of mine, Kent Quirk (formerly with Linden Labs, entrepreneur and game industry vet of some twenty years) had a stroke, and he used the Wii during his recovery. For him, it wasn't so much the activities the Wii allowed him to do that made difference—it was the ability to do them. I'll let Kent explain:

> Being able to stand up for ten minutes and play a game of *Wii Tennis* was a real victory for me. And I certainly wasn't going to go stand out on a tennis court and swing a real tennis racket at that point. So that was motivating and helpful to me. It was a simulation of real athletics, but at that point that was all I could handle. And even though I knew that, to feel like I could do that . . . it felt like a victory.

At the end of the day, the reality or unreality of the athletics didn't matter; it was success in the attempt that created a sense of empowerment and contributed to his recovery.

It's that feeling of empowerment—that you can beat the odds and come out on top—that ultimately makes health games like *Re-Mission* successful. Launched by the nonprofit HopeLab in 2006, *Re-Mission* had a very specific purpose. Richard Tate, HopeLab's Director of Communications, spoke with me about the game one afternoon. *Re-Mission*, he said, was created to address a real challenge for adolescents and young adults with cancer, who fare worse, in terms of outcome, than either adult or pediatric cancer patients—and it's related to a specific behavior: treatment adherence. A cancer diagnosis is frightening for anyone. For teens, though, who are already dealing with the normal changes to their bodies while trying to figure out who they are and navigate the terrain between child- and adulthood, the timing is especially difficult. Their lives are pulled apart, and they're being identified as sick—not what you want as a teen. As Richard told me,

> Oftentimes what happens is that kids go through really intensive initial treatment, and at a certain point they're being told by their doctors and their parents that they're in remission, they're cured. But they're also sent home with pills that they have to take on a daily basis, for sometimes up to two years. And those pills can be intensive in terms of their effects: they can make you break out, feel nauseous—and again, when you're trying to shift back into "I just want to be a normal kid" mode, the last thing you want to do is take these pills that are making you feel worse.

And this is where the problem arises: kids want to hang out with their friends and they don't want to feel nauseous or bloated, so they don't take their meds for a day. No big deal, right?

Actually, it is. Research shows that even missing a single day can have a disproportionate effect on their treatments. "That was the behavioral challenge," Richard said. "How can we create an experience that's going to get kids to stick to their meds?"

HopeLab had a couple of ideas around the psychological experience: One was knowledge: would kids fare better if they knew more about what they were fighting? The other was the psychological concept of self-efficacy: did they believe they *could* fight it? The question was how to reach these kids. They knew the status quo—handing them a pamphlet or DVD—wasn't going to cut it: the way kids interact with technology today is light-years beyond anything health communication can offer.

So they decided to create a game. Working with game designers, clinicians, doctors, nurses, scientists, cancer researchers, and most importantly the kids themselves, they built *Re-Mission*.

The game takes place in the year 2027, when scientists can now fight disease on the cellular level with intelligent nanobots. As the player, you control Roxxi, a nanobot designed specifically to fight cancer. You pilot her through twenty levels, designed around real experiences related to the common types of treatments in the most common types of adolescent cancer.

Before HopeLab released the game to the public, they conducted a thirty-four–site controlled study with 350 patients from across the United States, Canada, and Australia. What they found surprised them. "Kids who played *Re-Mission*," said Richard, "stuck to the treatments more consistently; they had greater self-efficacy and greater cancer knowledge."

That was in 2008. In 2010, HopeLab published a follow-up on the mechanisms of action. "What was it about the game," Richard said, "that was driving this behavior change? And we found that it was this interactive aspect of gameplay that was lighting up certain parts of the brain associated with motivation and reward and memory."

Perhaps most importantly, kids who played the game shifted their attitudes about their treatment, viewing chemotherapy as a weapon in their arsenal and a positive influence in their fight against this disease that had radically altered their lives. They were clear about what they needed to do and believed that they could do it. And in the fight to beat their cancer, that's more than half the battle.

If video games can help teach kids about what's happening in their bodies, the way they interact with them can also provide therapists a glimpse into the inner workings of their minds. As part of both his private practice and his work with Massachusetts General Hospital, Dr. Atilla Ceranoglu regularly uses off-the-shelf video games to help uncover issues that his patients might be unwilling or unable to express otherwise.

I met Atilla in his office in Milton, Massachusetts (just south of Boston). It's a very comfortable space: a few chairs, bookshelves, a variety of toys and games about, a bin of Legos, and an Xbox 360 hooked up to an LCD projector. We sat down to pastries and coffee (provided by the good doctor—since I did the driving, he thought he'd bring a little to eat), and eased into conversation—something that he does, not surprisingly, very comfortably.

Video games, he told me, are not central to his practice. He has them and uses them regularly, but doesn't push one mode of play over any other. He will on rare occasion make mention of a particular game he has in the office if some aspect of the game might be relevant to what a child is dealing with—along the lines of, "hey, have you heard of this game?"—and then he'll leave it at that. If the kid wants to play, great. If not, they'll move on to something else. It's all based on the patient. "Whatever they want to play," he told me, "I observe and engage through the play, and I try to make sense of what the child is going through or what the child is working through and how I can facilitate it."

Like the Legos or any of the other toys and games in his office, video games are just another tool that he uses to investigate a patient's psyche. I mentioned that I found it interesting that video games allowed a safe way for a child to express himself through the game character, removing himself a step from the issue. That was just play therapy, he told me. "There's nothing new about that. That's the art and approach in psychotherapy with kids, is when you're playing a game, keeping in mind that this is one degree removed." It's something you could do with a board game or other toy—checkers, for instance—which is an objection he's heard before (though he could just as easily respond that the kids he sees aren't necessarily playing checkers—which is often the case). The point, he says, is that video games are what many of these kids are playing, and they're sometimes the best way for a therapist to find common ground.

There *are* aspects of video games that are unique, though, and may provide opportunities for exploration and discussion not realized through other means. Video games are an immersive and highly visceral medium, and situations in gameplay can arise suddenly and be very intense. This may allow for a positive relationship between therapist and patient to develop quickly. Also, because of the emotional nature of play and the transference of identity from player to video game character (through the Proteus Effect), the sharing of experiences within the game may bring the patient and therapist closer together,[16] particularly if they're playing cooperatively. Competitive play in the safety of the therapist's environment may also help to build trust that the child might not find elsewhere.[17] So while video gaming is still play at heart, "it's a different form of play," he said, that "facilitates the child's expression of difficult emotional states."

Case in point: A few years ago, there was a young teen Atilla was working with, and they happened to be playing *Lego Star Wars*. They were wandering around the opening section of the game just generally blasting away at Lego people and breaking Lego structures, Atilla following behind his patient (who was playing the character of Princess Leia). Within this section of the game, there are several different doors that lead into the main game levels, and Atilla noticed that the boy would repeatedly head towards one of the doors, stop, and then turn away. "At first," Atilla told me, "I'm thinking, 'you know, this is not the first time he's playing,' I know, so he's clearly hesitant about going in. So I say something like, 'looks like Princess Leia has some second thought perhaps about going in there or not.' Something like that."

The kid's response was telling. "He says, 'well, we're gonna go in there and then we're gonna get shot, killed.' So very significantly, it's defeatist. Now, in that little . . . you know, it has a certain meaning, but when you compare it with what's happening outside, in parallel, now that has huge weight. What's happening there now gets more meaningful when you look at the child's life outside."

Outside, he was having trouble at home, at school, with friends. He was getting picked on and bullied. He had some learning disorders that made it difficult for him to succeed in a mainstream class, and he wasn't living up to his teachers' or parents' expectations—so he stopped trying and gradually developed this completely defeatist attitude. This was the source of Princess Leia's hesitation, and watching him play the video game gave Atilla the insight into this boy's mind he needed as well as the opening to explore and ultimately correct it.

The nature of video games also allows for exploration and even transference of roles from therapist to patient throughout their work together. For example, by winning a game against the therapist, the patient has the opportunity to adopt a position of power in the relationship, through consoling the therapist and/or responding to the defeat. Alternately, the therapist may be able to quickly take on the role of the patient and bring out other transference issues.[18]

Atilla related a story about a teenage boy he'd been playing *Halo* with in co-op mode (allowing them to work through the game together). Between sessions, the boy had been practicing at home, and he felt ready to take Atilla on head-to-head. Atilla beat him badly, they went on to play scoreless darts, and that was that . . . for a time.

> Meanwhile, he hooked it [his own Xbox] up at home, and then he came back and we played, and now he's beating me, and started to do well in the game. And at the same time, he was doing well outside, at school and all. . . . I was actively working with the school, actively working with the family. So, at the end, what happened was, as he's doing better, suddenly he started to have some challenges at school, and then he came back, and in the game, he's beating me bad, and I can't do much, and he says "are you letting me win?" So

> this is telling me that he can't take a defeat—he goes to play scoreless darts—but he can't accommodate a win either. He's not used to winning.

He and Atilla talked about it, and the boy admitted that he was used to having things get messed up. That was his default setting, so to speak, and he was unconsciously trying to return to that familiar ground. Through the gaming experience, though, he was able to articulate that to Atilla. "So next," he said, "what happened gradually is that we played *Halo* again, co-op, but now he's telling me where to go, he's taking charge, and that was the amazing thing."

For this boy, everything had come full circle—from playing co-op with his therapist to losing to him in head-to-head competition, practicing at home, finally beating him, trying to then sabotage a success he wasn't used to having, then finally being able to work with Atilla again, only this time taking on the role of team leader himself. "It was an amazing time," Atilla told me. "It was great."

But it was only a piece of the story. While working with the boy's parents, they identified an outside activity that he did well and that he loved: sailing. And at the age of fifteen, he took his parents out on a sailing trip with himself, as captain, at the helm—thanks, in no small part, to a video game.

There's another story I'd like to share with you, something I read just before this book went to press. It's from the video game blog *Hellmode*, founded by twenty-four–year–old Rhea Monique, who writes under the name Ashelia. She's an avid and experienced gamer, and she'd just finished playing Crystal Dynamics' brilliant and critically acclaimed reboot of *Tomb Raider*—the game franchise that launched heroine Lara Croft into the entertainment spotlight. For those of you who don't know Lara, think grittier female Indiana Jones and you'll get the picture. The original Lara hunted treasure, killed her enemies, and escaped perilous encounters all while sporting skimpy outfits and a gravity-defying anatomy—aspects that drew a fair amount of criticism from those who questioned why a strong female character had to look like a stereotypical teenage male fantasy in order to appeal to serious gamers (a critique I happen to agree with).

That was then. Thanks to the talented development team at Crystal Dynamics (some of whom are, in fact, women), Lara has been re-invented. Gone are the exaggerated figure and revealing attire, replaced instead by a tough, realistic young woman thrown into a ruthless environment who refuses to become a victim and, through strength, intelligence, and self-reliance, learns to survive a truly horrific situation.

And it is horrific. The game is rated mature for a reason: it is intense, it is violent, and it is gory. But it's not gratuitous. The violence is an accurate depiction of the harsh reality in which Lara finds herself, and the game would not be authentic or effective otherwise. Case in point: a scene early on when Lara, seeking refuge from the unfolding nightmare, is surprised by a man who lifts her up by the throat and begins choking her. As the player, you have mere seconds to press the right sequence of keys to save her before her eyes dim, her body goes limp, and she dies. It was at this moment that, for Ashelia, *Tomb Raider* became personal. On her first time through, she failed, and Lara's life was snuffed out in front of her. Ashelia was stunned. "I was so taken aback by the scene that I just stared at it, my mouth slightly open," she writes. "A video game had never made me feel this way in my entire life—and I wasn't sure what I thought about that."[19]

You see, twelve years prior, Ashelia survived this very scenario—at the hands of her own father. It was one of the last times she ever saw him.

It's not something she often talks about. She saw therapists and put the incident behind her. Got over it, moved on. And then Lara brought it all back, forcing her to relive an experience she'd spent most of her teenage years trying to forget.

Ashelia made several more attempts, dying repeatedly before finally hitting the right combination of buttons at the right time. She watched Lara kill her assailant.

And she cried.

The game had triggered something in her, had tapped into an emotion she'd driven deep. And by accurately portraying, in raw and brutal honesty, an event far too common for many women—and, more importantly, by Ashelia successfully navigating Lara through the game's many perils—*Tomb Raider* finally made her whole. She writes,

> It didn't hold any punches, but it didn't need to . . . it affected me in a way years of therapy never did. It healed me in a way that no one's physical comfort, words, and condolences could ever do. It made me realize that, much like Lara Croft, I survived as well—and that I had

> my own path to walk. That my experiences were real and tangible and yes, they defined me, but that I'd have it no other way. I am a survivor, and I am alive.[20]

After a dozen years of buried trauma and hidden pain, this young woman had found solace and salvation not with the help of another, but by her own hand, through Lara Croft, and through the terrible beauty of a game.

11

WAR GAMES

Combat Evolved

Your squad's on patrol when you get the call. A group of insurgents is causing trouble in a residential neighborhood just east of your position. There's no one else close enough to deal with them, so you're being sent in to clean up. You round up your men, give them the rundown, and head out.

As you approach the contact point, you hear the staccato burst of automatic weapons fire. A couple of blocks up, on your right. Close. You split your team up: half of you on one side of the street, half on the other. Moving cautiously. Checking corners.

Gunfire's louder, can't be more than a block away. Your pulse quickens, but you're cool. This is what you've trained for. Voices now, through the din. To a man, your squad drops into a crouch. You should be able to see them.

There. Just ahead. Eight, maybe ten men. Your squad moves in quarters now, two men advancing on one side of the street, two on the other, the rest watching their backs.

Half a block. You're almost on top of them. A shout, and something whizzes past your head, explodes into the wall next to you.

It's on.

Your team finds available cover, returning fire. The insurgents split up, ducking behind cars and into doorways. You drop one as he dashes into the open. Two more go down, then a third.

From across the street, a scream. One of your men has been hit. Two of his buddies lay down cover fire while the third moves him to safety. You tell your team to hold position, cover you, and then crouch-run to the other side. Halfway there, instinct tells you to drop and roll. Coming to your knee, you take down another insurgent who less than a second before had you pegged for dead. You check on your fallen comrade—he's been hit in the torso, but he's stable for now—and return to the fight.

Gunfire slows then stops as the last insurgent collapses in a heap. You check in with your corporal, and he confirms it: all insurgents down. One more of your men was hit, but not badly. He'll make it.

You stand up and are just about to give the all clear when a bullet slams into your chest. You stagger backwards. "Sergeant's down!" somebody shouts. "We need evac!"

And then everything goes dark.

Damn. Well, maybe next time.

You take off your headset and glance around the training room at the rest of your squad. "Okay sergeant," the trainer says, "take five and we'll try it again. And watch out for snipers this time."

• • •

In an actual firefight, trying again isn't an option: you either succeed or fail the first time—and failure can mean being taken out on a stretcher or going home in a body bag. Either way, it's not pretty. The US military invests heavily in training combat personnel to minimize casualties and ensure that they're as prepared as possible for the chaos of the battlefield and the harsh, unforgiving reality of war. Training isn't necessarily a simple proposition, though. There are multiple methods, and choosing the most effective one requires balancing several variables, including intent, budget, space, time, and the number of people involved.

Broadly, military training falls into three categories: live, constructive, and virtual. Live training is conducted in an actual, physical location set up for a specific purpose (urban warfare, desert combat, etc.). It's very effective, but it's expensive to organize and run—and gets more so the more people you have to train—and isn't very customizable. Constructive training takes place, again, in a physical environment. It has elements of live training and also uses computers to simulate elements of combat. It's less expensive and easier to customize, but it has a lot of moving parts to coordinate, and it relies in some measure on computer simulations, which vary in quality and degree of realism. The third method is virtual training, which takes place entirely within a virtual environment and is completely computer-based. Virtual training is the least expensive and most flexible of the three, but its effectiveness can diverge greatly, again depending on the fidelity of the simulation (not just visual, but in terms of the nature of the content).

Historically, high-end simulations were restricted to the realm of aviation—helicopters and airplanes being rather costly to replace[1]—and they've been very successful. Even basic aviation-themed sims, though, have had excellent results: Army pilots who trained on Microsoft *Flight Simulator* scored better on the aeronautical segments of their training than those who didn't. To Colonel Scott Lambert (ret.), formerly of the Program Executive Office for Simulation, Training and Instrumentation, it makes perfect sense. "I think what they found is the people who could understand the rules of aeronautics in the context of the behavior of the plane did better because it had more meaning to them. It became a little more intuitive."

Infantry training has, for the most part, fallen under the purview of live simulations—often on an epic scale. When Lambert was on his tour of duty in Europe, the Army conducted an exercise called Return of Forces to Germany, during which it deployed the entire corps, a Herculean undertaking. Perhaps the ultimate example of a large-scale simulation is the National Training Center in Fort Irwin, California, modeled during the Cold War on meeting Soviet forces in battle on the plains of Germany. "The training area is almost the size of Rhode Island," said Lambert, "and you would deploy brigade-sized units—armor and artillery and everything." Executing these exercises, he said, cost billions. And each time you set one up, you could only run it once.

Budgets for training of this magnitude just don't exist any longer, or at least not to the extent they once did. At the same time, the need for effective training is probably greater now than it's ever been, given the diverse and complex situations our military is faced with. In this environment, technology is taking on an increasingly important role. "The demand for simulated solutions to training is only increasing," said Lambert. "But the good news there is the technology is only getting better at much, much lower cost."

Case in point: *Virtual Battlefield Simulator 2 (VBS 2)*, a game-based battle simulator first purchased by the US Marine Corps in 2001, regularly updated and still in use today. Developed by Czech games company Bohemia Interactive, *VBS 2* runs on a standard laptop computer and allows soldiers to work alongside each other in a virtual environment that's equivalent to hundreds of square kilometers.[2] The simulator features weapons that function realistically, a variety of land, sea, and air vehicles that users can operate, and an environment that commanding officers can adapt as needed (adding rain or fog, for example). *VBS 2* is built on an open platform, which means that soldiers can update its database with data from their own experiences in the field; the latest version of the simulator can also import detailed geographic and satellite images, allowing users to create virtual versions of actual locations and then repeatedly rehearse them.[3] According to First Lt. Roy Fish, a platoon commander in Afghanistan, *VBS 2* has clearly saved lives. "Every time we go outside the wire and react to an IED [improvised explosive device] or small-arms fire, it all translates to what we did in the simulator."[4] Adding to its value as a training platform, *VBS 2* includes an after-action review component that records every action, explosion, environmental factor, and vehicle movement. Teams can review footage of their performance to see what went right and what went wrong—and hopefully identify what they can do to correct it.[5] In Army Colonel Anthony D. Krogh's opinion, *VBS 2* is arguably "one of the most successful simulations we've ever brought to the force. In terms of cost versus usage, it's a huge success."[6]

Technology seldom stands still, though, and as good as *VBS 2* is, the end of 2012 saw the launch of the sim that would be king: the Dismounted Soldier Training System (DSTS), the most advanced virtual reality simulation ever built. The DSTS is the first completely immersive, 3D virtual training system, and it represents nothing less than a fundamental and complete transformation in the way service personnel train for combat operations.[7]

Developed by Intelligent Decisions, based in Ashburn, Virginia, and built on the groundbreaking CryENGINE® 3 game engine (first used in Electronic Arts' 2011 game *Crysis 2*), DSTS is a 360-degree virtual combat zone. Up to nine soldiers (represented by individual avatars) can enter DSTS at once, via standard-issue combat helmets outfitted with virtual reality visors, stereo speakers, a head-tracking system, and a microphone. A lightweight computer carried on each soldier's back processes the virtual environment that s/he sees through the visor, updating and displaying any changes to the setting instantly. The helmet's head-tracking system combined with sensors worn on the body register a soldier's every movement precisely—allowing for completely natural motion—and translate it to the avatar within Dismounted Soldier's world: if a soldier ducks, the avatar ducks. If s/he rolls, the avatar rolls. Soldiers communicate with each other using the helmet's microphone; the speakers transmit voice and full surround sound, completely immersing them in the symphony of war. Each soldier also carries a fully functional replica of his or her actual weapon; it fires virtual bullets, but operates like a real weapon in all other respects (including the need to be reloaded). A soldier's avatar can also be injured within the DSTS and the injuries take accurate toll on the virtual body: suffer too many hits and it's game over. The beauty of a virtual training space is that death's not final: soldiers can try strategies that may be risky or dangerous in the real world, fail safely, and start over.

Along with capturing and transmitting each soldier's movements, DSTS can recreate the movement of land, water, and air vehicles and guided weapons, display a variety of landscapes and details, and dynamically update changes to the virtual environment—from explosions to footprints—providing an incredibly heightened sense of realism. "What we're trying to do," said Chris Stellwag of CAE, ". . . is to be able to change the virtual world dynamically, so when things happen in real time, the virtual world changes."[8]

Along with the expanded scope and reach provided by virtual environments like DSTS and *Virtual Battlefield Simulator*, commanding officers and strategists have an unprecedented level of power: they can replicate an exact real-world location or building—down to carpeting and furniture in the lobby—and train combat troops for an unlimited number of scenarios. It's something that Rob Lindeman of Worcester Polytechnic Institute is very familiar with.

> In the game world, you can set it up to be anything. So we get some hostage rescue work where there's an embassy somewhere, and they're holding some people hostage and they're gonna send in a strike team of four to go in and rescue the hostages. So we can actually produce an environment that has the exact floor plan of that embassy so they can actually train in that space. They can look

> around and get a feel, so that when they actually go, there are fewer surprises.

The value of this shouldn't be underestimated. With a highly accurate simulation in their hands, your strike team would know the exact layout of the building—sight lines, stairwells, crawl spaces, entrances, and exits—they'd know the minimum number of people needed to take it, and they'd know exactly where to place them to be most effective. And they'd know all this not from looking at construction plans or photos but because they'd been through it before. In fact, there's some speculation that this has been done already. An unnamed source very familiar with the military training business and quoted in a special section of *The Economist* certainly thinks so. "It's quite likely that the team that killed Osama bin Laden would have rehearsed the raid in some sort of virtual environment."[9]

As valuable as virtual reality sims are in training our combat personnel for the exigencies of deployment, there is a more laudable purpose they can serve: taking care of them when they return home.

Picture this: you're behind the wheel of a military Humvee on the road to Fallujah, your corporal's in the seat next to you, a fire team of Marines is in the back. Tensions are high: Iraq is still a hotbed of violence, you're traveling a dangerous road, and everyone knows the risks. Still, nothing's happened yet. You're just beginning to relax when a roadside bomb—one of Iraq's infamous IEDs—rips through the truck with a deafening roar. The corporal dies instantly, but you barely have time to notice because the Humvee's now on its back. Screams sound from behind you. Looking back, you can see your team through the billowing smoke—and it's not pretty. A Hollywood makeup artist with an unlimited budget and a taste for the macabre would have a hard time duplicating the scene. Some of the men are dead, the rest horribly wounded. There's something burning in the back, and the noise and smoke are overwhelming. You have to get out, but shock's taken over and you can't move.

I sincerely hope that none of you have ever had such an experience, or know anyone who has. Unfortunately, for many US servicemen and women, variations on this scene are all too real. And for those who survive, healing from the physical wounds may be the easy part.

Post-traumatic stress disorder, or PTSD, has always been a serious problem, and it's getting worse. Iraq and Afghanistan are unique in the history of US military conflict (in terms of the length of deployments and the speed at which troops are redeployed) and seem to be contributing to growing mental health problems: according to a RAND Corporation study, close to 20 percent of Iraq and Afghanistan veterans suffer from PTSD or major depression.[10] According to Dr. Steven Huberman, dean of Touro College's School of Social Work in New York City, "Since the deployment to Iraq and Afghanistan started . . . we're seeing a significant difference from other military involvements, in the number and types of injuries, the types of deployments, the nature of the military force, and the impact on families and kids."[11]

PTSD is often hard to identify, always difficult to treat, and has far-reaching impacts on sufferers and their families. In order to recover, victims must confront the memories and emotions surrounding the traumatic event and eventually work through them. Ignoring them only creates more severe problems. Dr. Michael Kramer, a psychologist at Manhattan's Veterans Administration Hospital, uses exposure therapy—asking patients to face traumatic memories head-on and describe them in as much detail as possible—to help PTSD sufferers confront and eventually overcome their trauma. It can be painful, but it works: Once the memories are out in the open, therapists can help their patients separate the memory from the emotional trigger that sets them off. "One of the hallmarks of PTSD is avoidance," said Dr. Kramer. "Patients spend an awful lot of time and energy trying not to think about it or talk about it. But behaviorally, avoidance is what keeps the trauma alive."[12] The trick is confronting the memories safely.

So how does he accomplish this? By using a tool that combines sight, sound, touch, and smell into a single engaging experience: virtual reality.

Enter *Virtual Iraq*. *Virtual Iraq* is an immersive, 3D virtual world that allows a patient with PTSD to relive a traumatic situation in a safe environment. Based on the video game *Full Spectrum Warrior*, *Virtual Iraq* places the patient into a therapist-controlled combat scenario. During the scenario, the therapist exposes the veteran to the sights and sounds of battle at a level that s/he is emotionally capable of handling (it's known as virtual reality exposure therapy). As the patient progresses, the therapist can turn up the heat, enhancing the realism of the scene by delivering additional sounds and images—jets flying over, insurgents coming out of palm groves, IEDs, explosions, a Muslim call to prayer—into the environment. The video game provides a safe way for the patient to confront his or her emotions and ultimately gain control over the PTSD.

Says Albert "Skip" Rizzo, clinical psychologist, associate director for Medical Virtual Reality at the University of Southern California's Institute for Creative Technologies (ICT) and *Virtual Iraq*'s developer,

> [Virtual reality] puts a person back into the sights, sounds, smells, feelings of the scene. . . . You know what the patient's seeing, and you can help prompt them through the experience in a very safe and supportive fashion. As you go through the therapy, the patient may be invited to turn on the motor. Eventually, as they tell their story, you find out that it wasn't just a vehicle in front, it was a vehicle with five other friends. . . . The guy that died was going to be discharged in two months. You start to see a rich depth of story.[13]

Okay, all well and good. But does it work? This is still a fairly new area, and Rizzo is careful not to draw too many conclusions from early results. However, using accepted and well-established diagnostic measures, he said, about 80 percent of people who've completed treatment showed both "statistically and clinically meaningful reductions in PTSD, anxiety, and depression symptoms, and anecdotal evidence from patient reports suggested that they saw improvements in their everyday life situations. These improvements were also maintained at three-month post-treatment follow-up."[14]

This is huge. PTSD is pernicious and notoriously challenging to treat. If this proves successful on a large scale, Rizzo will have found the ideal solution, one that has no side effects, carries no risk of dependency, and can be administered for as long as is necessary.

Perhaps the best testament to the effectiveness of *Virtual Iraq*, though, comes from this twenty-two–year–old Marine injured during combat operations in Iraq. This was after six weeks of virtual reality exposure therapy:

> Before, I felt like there were two people in me, the marine, who was numb, who was a tough guy, and the civilian me, the real me, the guy who isn't serious all the time, the guy who can take a joke. By the end

> of therapy, I felt more like one person. Toward the end, it was pretty easy to talk about what had happened over there. We went over all the hot spots in succession. I could talk about it without breaking down. I wasn't holding anything back. I felt like the weight of the world had been lifted. I was ready to be done. The last two sessions, I didn't think I needed to be there anymore.[15]

This young man—and there are others like him—gained his life back in large part through the healing power of a video game.

Results like this are certainly encouraging, especially if they're borne out in further studies. Given what we're asking these kids to sacrifice and expecting them to endure, the least we can offer is the hope that they may someday regain their lives. Wouldn't it be better, though, if we could prevent PTSD from taking hold at all? Is that even possible?

The answer to both those questions is, perhaps surprisingly, yes. And it's an idea that Rizzo and his team at ICT—with funding from the US military—are aggressively pursuing. Their goal? To drive PTSD to extinction. They hope to prepare military personnel for the psychological trauma of war before they've even been deployed and stop PTSD from ever finding a foothold.[16]

To this end, Rizzo is gathering stories from returning veterans and recreating their experiences in virtual reality, building simulations that he can then use to, in essence, protect the minds of troops heading into war zones. "What we want to create," he said, "is something that pulls at the hearts of people. Maybe there's a child lying there with the arms blown off, screaming and crying. Maybe your action kills an innocent civilian, or you see a guy next to you get shot in the eye with blood spurting out of his face."[17]

At that point, Rizzo's intent is to freeze the simulation and have a virtual character—someone perhaps specific to the trainee who helps him or her keep calm in the face of virtual trauma—walk the trainee through the rest of the scenario. Rizzo's thinking is that, if the trainee learns coping skills when emotionally aroused, s/he'll retain them better and be more likely to draw on them if s/he ends up in a similar state in real life. The military's used this type of stress-resilience training for a couple of years already. The difference is that a virtual reality simulation, or even a game, could extend the reach of this kind of training to more and more people.

"Nobody goes to war and comes back the same," Rizzo said, "but when they return are they capable of holding a job and loving their wife and kids? That's what our aim is, to make the return home as smooth as possible."[18]

Of course, there's another way this could go. We have, at our disposal, a tool that may help us reach some of our military objectives while, at the same time, reduce the number of our enlisted men and women in harm's way. Unmanned aerial vehicles. UAVs.

Drones.

Depending on who you ask, public opinion about drones ranges from them being, at one end of the spectrum, a panacea for our overstretched military to, at the other extreme, a sure sign of the decline of American morality and our inevitable slip into militaristic totalitarianism. As with most complicated issues, the truth probably lies somewhere in the middle—though the issues arise, I believe, more from their use than their reality.

Drones do provide some unquestionable benefits. They allow us to operate in military hot zones—even if only in a surveillance capacity—without putting our own troops at risk. They allow us to reduce boots on the ground, freeing up resources for other areas. And when used well and conscientiously, they're a more precise military tool than traditional air strikes or bombardment. However, the very aspect of these drones that makes them beneficial—they're unmanned—also makes them ethically problematic.

From the beginning of time, the object of warfare has been to dehumanize the enemy and therefore make them easier to kill. After all, you're no longer taking a human life—or if you are, they're somehow not quite as human as us. The progress of military technology (if you can call it that) has meshed well with this philosophical practice, allowing us to kill at greater and greater distances. We started with bare hands, clubs, knives, and swords. Killing another human being was an intimate personal affair: you could feel, hear, and smell your enemy. It was impossible on a visceral level to deny that person's humanity. The advent of firearms allowed us to take a step away. You still had to be in sight of your enemy, but you lost some of that human connection. Aerial bombing pulled us even farther back. Now, you were just targeting buildings, with no evidence of the human life within. However, you were still there, and there was still potential risk of harm.

And now we come to the age of drones. War by remote, carried to the extreme. A drone pilot can commute to work from a DC suburb, put in an eight-hour shift, maybe bomb the hell out of some village half a world away, and still make it home in time to catch his or her kid's soccer match. All chance for human connection with your target has vanished. How hard is it to dehumanize the enemy now, when you're not even on the same continent? It's like a game, especially when you consider that the people at the controls grew up playing games, and they're now piloting weapons of destruction with video game–style controllers and viewing the world through what amounts to a video game screen. "Drone operators kill at the touch of a button," writes Michael Hastings, "without ever leaving their base—a remove that only serves to further desensitize the taking of human life."[19] Surely we must be trivializing war, sanitizing it. Certainly making it easier to carry out. Though exact numbers are hard to come by (the CIA has a covert drone program that it just does not talk about), it seems the Pentagon has around nineteen thousand drones in service right now.[20] I suspect it's far greater. If it is the case that we're making it easier to engage in warfare, and we have tens of thousands of drones in operation right now, then I'm frightened.

When I started looking into this, I had a very definite perspective. Of course this was dangerous. Piloting a powerful weapon using game controllers, viewing the world through a game-like interface, how could those in control not mistake this for a game? Many people I spoke with had similar concerns.

I was very surprised, then, to learn that in general, the actual pilots look at this as anything but a game. In fact, in story after story, drone pilots (they prefer remotely piloted aircraft, or RPA) work very hard to make the rest of us understand exactly the opposite. According to Joshua,[21] former sensor operator (he controlled the RPA's camera) and now trainer at Holloman Air Force Base in New Mexico, "One of the things we try to beat into our crews is that this is a real aircraft with a real human component, and whatever decisions you make, good or bad, there's going to be actual consequences."[22]

One of the features of these drones oft-cited as turning them into video games is their surveillance cameras. The truth, it turns out, is a bit more complicated. These cameras are powerful, bringing the war very much into view for the pilots—who often spend days or weeks getting to know the habits of a particular target—a militant, for example—watching him interact with his family, kids, friends, and neighbors. They try to time their strike when his family is out, but while they're waiting, they get to know him. "They watch this guy do bad things and then his regular old life things," said Colonel Hernando Ortega, chief of aerospace medicine for the Air Education Training Command. "At some point . . . you might gain a level of familiarity that makes it a little difficult to pull the trigger."[23]

Difficult indeed. And when they do pull the trigger, that often sets another difficult situation in motion. From the safety of his computer desk in the suburbs of Syracuse, New York, Colonel D. Scott Brenton pilots a drone that patrols the skies of Afghanistan. There's a fundamental disconnect between the war he's helping fight and the location from which he's fighting it—a disconnect that grows more profound when he's called upon to pull the trigger. "It's a strange feeling," he said. "No one in my immediate environment is aware of anything that occurred."[24] Instead, he starts the commute home, re-entering the world around him, yet separated from it on a basic, human level, alone in the knowledge of his actions.

Far from reducing their jobs to a glorified video game, RPA pilots seem painfully aware of the reality of the death they deliver. And though they can often justify it, they never forget it. Major Bryan Callahan summed it up well. "We're well aware that if you push that button, somebody can go away," he said. "It's not a video game. You take it very seriously. It's by far nowhere near a video game."[25]

What seemed simple to me at the outset has shifted. Far from the casual killers I assumed I'd find, instead I discovered thoughtful human beings faced with responsibilities and decisions the likes of which most of us will never know. It seems I was wrong. The technology that made it logistically easy for these men to kill remotely only made it more important for them to understand the gravity of the act.

Michael Hastings wrote a brilliant and thorough exposé of drones for *Rolling Stone* magazine. Early on, he illuminates the central issue. On a broader scale, he says,

> The remote-control nature of unmanned missions enables politicians to wage war while claiming we're not at war—as the United States is currently doing in Pakistan. What's more, the Pentagon and the CIA can now launch military strikes or order assassinations without put-

> ting a single boot on the ground—and without worrying about a public backlash over US soldiers coming home in body bags. The immediacy and secrecy of drones make it easier than ever for leaders to unleash America's military might—and harder than ever to evaluate the consequences of such clandestine attacks.[26]

Drones are powerful, fantastic, terrifying tools. They can do great things and they can bring about unspeakable horrors. Going forward, we have to think very carefully about how we use these tools, and, even more importantly, why. And as we rush to embrace this new technology, we must consider the consequences—to us as a nation, to the intended victims when our attacks succeed and the unfortunate tragedies we create when they fail, and to those whom we ask to act on our behalf, and for whom the terrible brutality of war may be distant, but not forgotten.

12

IT'S WILLIAM GIBSON'S WORLD, WE'RE JUST LIVING IN IT

Ice walls flick away like supersonic butterflies made of shade. Beyond them, the matrix's illusion of infinite space. It's like watching a tape of a prefab building going up; only the tape's reversed and run at high speed, and these walls are torn wings.

. . . The core data tower around us like vertical freight trains, color-coded for access. Bright primaries, impossibly bright in that transparent void, linked by countless horizontals in nursery blues and pinks.

But ice still shadows something at the center of it all: the heart of all Chrome's expensive darkness, the very heart. . . .[1]

—William Gibson, "Burning Chrome"

The year is 1981. An audience of four listens as William Gibson spins "Burning Chrome"—a tale of console cowboys, cyberspace, and virtual reality set in a future not far off. Three years later, *Neuromancer* introduces cyberspace to the rest of the world, simultaneously reinventing science fiction and challenging our concept of reality.

Over the next two decades, Gibson continues to present us with his dark visions of a society ruled by technology and governed by information. And though his dystopian future has not (yet) come to pass, many of his ideas are now commonplace—cyberculture, neural implants, reality television, and the ubiquity of the Internet among them. If you've never read Gibson, but have seen *The Matrix*, *Vanilla Sky*, or *Lawn-*

mower Man, or been pulled into a video game's virtual environment, then you've been under his spell.

I first discovered Gibson in a literary journalism class at Hampshire College in the late '80s—through "Burning Chrome," in fact. It was one of the first stories our professor, Norm Sims, had us read, and from the opening paragraph, I was hooked. Gibson's style was knife-edged, visceral, cinematic, his vision of the future dangerously seductive. Reading "Burning Chrome" was like dancing across the dark edge of humanity. The story's twenty-three pages worked like a gateway drug, compelling me to take more. I devoured his short fiction—"Johnny Mnemonic," "Hinterlands," "Fragments of a Hologram Rose"—working my way up to larger doses. In *Neuromancer*, I fell in love with his mirror-eyed antiheroine, Molly Millions; in *Virtual Light*, it was the Bridge. But what really drove me to seek out Gibson in whatever form I could find was cyberspace—a dream-shifted realm where formless data takes shape in glowing towers of light, guarded by digital security protocols set to devour the mind of any console cowboy foolish enough to jack in and steal it.

As with the best science fiction, Gibson's stories have either predicted or played midwife to the birth of much that we take for granted. I mentioned a few examples here, but there are many others. Nina Fefferman (the Rutgers epidemiologist) reminded me that his ideas are credited as a driving force behind the inception of *Second Life*. For better or worse, cyber-crime owes a debt to him. And two pieces of emerging technology were clearly inspired, at least in part, by his work.

The first you've probably heard of and may even own: augmented reality (AR) glasses. AR glasses, typified by Google Glass, are lightweight goggles that connect to the Internet via hands-free wi-fi and allow the wearer to simultaneously interact with the real and virtual worlds. In the case of Google Glass,[2] there's a rectangular window in the upper-right field of view that serves as your interface with the 'net. The glasses have a built-in microphone, camera, and headphones, allowing you to transmit audio, real-time video, and still photographs over the Internet. If you choose to, you can allow anyone on the planet to see the world through your eyes.

Unless you're enlisted in the armed forces or part of the federal government, you may have to wait awhile for the second piece of technological wizardry. Taking a page out of Gibson's *Virtual Light*, the US

Department of Defense is developing virtual reality contact lenses to enhance the intelligence, surveillance, and reconnaissance abilities of soldiers on the battlefield. The lenses, which fit over the eye exactly like standard contact lenses, contain miniature, full-color displays on which digital images can be directly projected. Wearers could watch these images and still have an unobstructed view of their surroundings, allowing them to react to real-world events while receiving potentially critical data through the lenses in real time.[3] The good news for us civilians is that, unlike Vegas, what happens in the military doesn't stay there; once they're rolled out in the name of national security, it won't take long for someone to adapt these lenses for the general consumer, or develop a set of AR glasses that provide the same functionality.

Of course, there are some things Gibson *didn't* predict—many of which are already available. Consider the Kinect, for example. This $150 motion-capture camera transformed the world of video gaming by allowing gamers to control some of their favorite titles using only their bodies. However, it quickly expanded far beyond its original purpose: a University of California, Berkeley, group attached a Kinect to a quadrotor (a helicopter with four propellers) and got it to fly around a room on its own; students at Munich's University of Bundeswehr added one to a robotic car, enabling it to successfully navigate an obstacle course; and two students at the University of Konstanz in Germany used the Kinect to create a navigational system for the blind.[4]

In the realm of mind and body, several companies—most notably EmSense, Emotiv, MindGames, and NeuroSky—are developing technology that detects brainwaves, allowing users to control video games with their minds.[5] This technology has applications far beyond entertainment: Both Honda and Toyota are investing a lot of green into researching mind control features, such as trunks and doors that open by thought command. The Defense Advanced Research Projects Agency has even gotten into the act; it awarded Johns Hopkins University a cool $34.5 million in 2010 to test mind-controlled prosthetic limbs.[6] And lest we forget the body, a new technology called Surround Haptics, developed at Disney Research, Pittsburgh, allows gamers to feel a wide range of sensations—from gentle caresses to bone-jarring collisions.[7] Initially demonstrated during the Emerging Technology Exhibition at SIGGRAPH 2011, it was used to enhance a driving simulator, allowing drivers to feel road imperfections, skidding, braking, acceleration, collisions, jumping and landing, even objects striking the car.

Clearly, there's a lot of energy going into hardware development—the *how* piece of the perception equation. But that's only half the story. What about the other piece, *what* we perceive? Without improvements in the quality of the content, all the fancy hardware in the world will be wasted. What's happening in this arena?

When you start talking about enhancing what we see, there are four main issues. The first, of course, is artistic—creating high-quality, photorealistic animation. The other three are technical—rendering the artwork in real time, ensuring that it behaves consistently with the real world (the physics model), and finally, having the bandwidth to deliver the content, again in real-time, over the Internet. Industrial Light and Magic's Kim Libreri has some insight on these matters—and he's optimistic on all of them. Regarding image quality, he told me that we'd see absolute photorealism within five years. By that time, he said, "Real-time computer graphics are gonna be indistinguishable from reality."

That's the first—and easiest—hurdle to leap. In the short-term, this will affect video games or other programs running locally—in other words, on an individual game console or computer, or, at most, a local area network of machines hooked directly up to each other. To achieve absolute photorealism in an immersive virtual environment requires the ability to do real-time, hi-fidelity rendering and employ real-time physics systems that accurately express factors like gravity and explosions.[8] We'll be past both of those obstacles, Libreri told me, within a decade at most. Given that Industrial Light and Magic is rendering *Star Wars 1313* at a million times faster, per frame, than ten years ago—a pace of change that far exceeds Moore's Law—it's not hard to see this happening. All that leaves is the network infrastructure—the Internet bottleneck, if you will. If you can improve the Internet technology enough to support the rendering and physics engines, he said, then why not run a single game across multiple servers? "You could have a game that is running on ten computers at the same time," he said, "or one hundred computers, or one thousand computers . . . you could have massively parallel physics simulators or graphical rendering. You could render it on a separate machine and then it's real-time composited together on some big, crazy Google or Amazon [server] farm."

According to Libreri, all conventional wisdom points to the Internet bottleneck being cleared within a decade as well.

And then that's it. We will have cleared the last obstacles away from both parts of the perception equation—the *how* (technology) and the *what* (content)—and removed the last barrier to delivering hyper-realistic, fully immersive virtual worlds. And once AR glasses become ubiquitous (and if our history with smartphones and tablets is any indication, they will), we'll be able to experience those worlds wherever we are. It's a safe bet that, sometime in the not-too-distant future, many of us will spend our days working either entirely within a virtual world or bouncing back and forth between the virtual and the real throughout our day.

If that becomes the case, it could lead to a curious and potentially troubling circumstance. "We're a very flexible learning device—the human brain is," said Libreri. "With augmented reality and the fidelity that we'll get over the next decade, I think it's gonna be quite hard for people to separate what's real and what's not real as we evolve as a race."

The longer we coexist within those places, the harder it will become, and this could begin to create problems. Take the area of cyber-security. Cyber-security's a concern now. How does that evolve when we're all walking around with AR glasses? "There's a real, tangible threat to what can happen to you in the cyber world," said Libreri, "and I think as things visualize themselves more realistically . . . think about cyber-crime in an AR world. Creatures, like, chasing you. It's gonna be pretty freaky. People will have emotional reactions to things that don't really exist."

It might also render people particularly susceptible to suggestion. It's a bit like the movie *Inception*, where they're planting ideas deep within a subject's subconscious, except this would be easier. If you can't distinguish the virtual from the real, a suggestion made to you in a virtual world could seem like a really good idea someone gave to you in reality.

As intertwined as they are, in this scenario there's still a recognizable, if tenuous, distinction between virtual and real. So what's left?

Tapping the brain, of course.

Ever since human beings invented the motion picture, the history of the field's technological development has been one of bringing the viewing platform closer to the brain, shortening the path from stimulus to perception to brain activation. Starting with movie screens, the path looks like this (and we could draw a similar path for audio stimulus):

> movie screen → eye surface (dozens of feet) → retina → optic nerve → brain

The paths from home television to brain and tablet to brain are similar, with progressively smaller distances between the screen and the eye. The next major leap was virtual reality goggles/contact lenses. That path looks like this:

> virtual reality glasses/lenses → eye surface (inches away to touching) → retina → optic nerve → brain

Though it's not widely discussed yet, we now have the ability to, using a nonburning laser, draw directly onto the retina.[9] We've shortened the path considerably:

> retina → optic nerve → brain

We've just eliminated two entire sections. At some point in the future, we'll tap into the optic nerve itself and from there we'll eventually figure out how to stimulate the areas of the brain that correspond to our senses directly, eliminating the need for any external input device. Once you cross the threshold to direct brain stimulation, the virtual world will become, in any way that matters, completely real.

What seemed far-fetched and confined to the realm of science fiction less than thirty years ago is all but upon us. In another thirty years, we may be able to live much of our lives in a completely virtual world that's indistinguishable from reality. Whether this turns out to be a boon or a curse will be debated long after its inauguration. However, the history of human technological progress has taught us that there's no turning back: Once we have the means to create something, it's a virtual certainty that we will.

Unless, that is, we already have.

Suppose, just for a moment, that none of this is real, that we are, all of us, inside a simulation, a game. How would we know? What we define as real is only the result of electrical signals interpreted by our brain. Based on that measure, would the game be any less real than reality?

As human beings, we are the sum total of our life's experiences. Whether they transpire within the real world or a simulated representation, their actuality is unquestionable. They shape us into who we are and help guide us towards who we will become.

As human beings, we are also gamers. The games we create and the games we play tell us something about our society and ourselves: what we desire, what we value, what we encourage, what we fear. At their best, they can open our eyes, expand our minds, unite us to common purpose, and drive positive change. However, they also present a challenge: as they approach reality, there is a danger that we may retreat inside ourselves or seek constant refuge from the bitter aspects of the real, eschewing reality for the game.

Will we ever forget that we're playing? Perhaps. If not, games will simply evolve along with us, entertaining us, educating us, and reflecting the dark and light of our humanity. If so, we may find ourselves imprisoned in a virtual cage of our own making, reality fading into a dream while the outside world carries on or falls into ruin in spite of us. As with everything, the future of video games is rich with opportunity, but the way ahead lies shrouded in the unknown. We can guide their evolution in whatever direction we choose, but we must proceed, as in life, with caution and with care, conscious of our checkered history and mindful of the consequences our choices, and our actions, may beget.

To some degree, life itself is a game. We don't know all the rules, it doesn't always turn in our favor, the choices we make determine its course, and the outcome, though certain, isn't always expected. It can be beautiful, and it can be tragic. There are great triumphs and bitter defeats. There is pain, happiness, sorrow, and joy. At times, we struggle to continue, to find the inner strength to persevere, to carry on, to keep playing. And yet we choose to keep playing.

Because in this game, we are not alone.

Because in this game, we're playing for more than ourselves.

And because through this game, we declare to the universe that we are gloriously, wondrously alive, and that we, in our fleeting tenure, made a difference.

NOTES

1. DOWN THE RABBIT HOLE

1. The Entertainment Software Association, *2012 Essential Facts About the Computer and Video Game Industry* (The Entertainment Software Association, 2012), 11, accessed September 5, 2012, http://www.theesa.com/facts/pdfs/ESA_EF_2012.pdf.

2. Motion Picture Association of America, *2011 Theatrical Market Statistics* (Motion Picture Association of America, 2012), 2, accessed September 5, 2012, http://www.mpaa.org/Resources/5bec4ac9-a95e-443b-987b-bff6fb5455a9.pdf.

3. WR Hambrecht+Co, *The U.S. Professional Sports Market & Franchise Value Report, 2012* (San Francisco: WR Hambrecht+Co, 2012), 22, accessed September 5, 2012, http://www.wrhambrecht.com/pdf/SportsMarketReport_2012.

4. "New Video Game Fights Teenage Depression," *New York Daily News*, August 6, 2012, accessed September 5, 2012, http://articles.nydailynews.com/2012-08-06/news/33069746_1_teenage-depression-depression-in-young-people-adolescent-depression.

5. Constance Steinkuehler and Sean Duncan, "Scientific Habits of Mind in Virtual Worlds," *Journal of Science Education and Technology* 17, no. 6 (2008): 530–543.

6. Bradley Bouzane, "First-ever Virtual Surgery Performed," *Winnipeg Free Press*, August 21, 2009: A18.

7. Steve LeBlance, "Studies: Video Games can Make Better Students, Surgeons," *USA Today*, August 19, 2008, accessed September 5, 2012, http://www.usatoday.com/tech/gaming/2008-08-18-video-games-learning_N.htm.

8. Johan Huizinga, *Homo Ludens: A Study of the Play Element in Culture* (Boston: The Beacon Press, 1955), 9.

9. *Ibid.*, 101.

10. Stephen P. Siwek, *Video Games in the 21st Century: The 2010 Report* (The Entertainment Software Association, 2010), 27, accessed September 5, 2012, http://www.theesa.com/facts/pdfs/VideoGames21stCentury_2010.pdf.

11. The Entertainment Software Association, *2011 Essential Facts About the Computer and Video Game Industry* (The Entertainment Software Association, 2011), 11, accessed September 5, 2012, http://www.theesa.com/facts/pdfs/ESA_EF_2011; The Entertainment Software Association, *2012 Essential Facts About the Computer and Video Game Industry* (The Entertainment Software Association, 2012), 11, accessed September 5, 2012, http://www.theesa.com/facts/pdfs/ESA_EF_2012.pdf.

2. FROM THE COIN-OP TO THE CONSOLE

1. Steven L. Kent, *The Ultimate History of Video Games: The Story Behind the Craze that Touched our Lives and Changed the World* (New York: Three Rivers Press, 2001), 23–24.

2. In a move to try and invalidate Baer's patents (and thus stop paying royalties to Sanders Associates), Nintendo sued Magnavox in 1985, claiming that Willy Higinbotham had actually invented the video game with *Tennis For Two*. The court ruled in favor of Baer and Magnavox, writing that *Tennis For Two* didn't use video signals, and thus was not a true video game. Baer's patents stand to this day.

3. J.M. Graetz, "The Origin of Spacewar," accessed September 24, 2012, http://www.wheels.org/spacewar/creative/SpacewarOrigin.html.

4. J.C. Herz, *Joystick Nation: How Videogames Ate Our Quarters, Won Our Hearts, and Rewired Our Minds* (Boston: Little, Brown, and Co., 1997), 7.

5. Nolan Bushnell, quoted in Steven L. Kent, *The Ultimate History of Video Games: The Story Behind the Craze that Touched our Lives and Changed the World* (New York: Three Rivers Press, 2001), 34.

6. Ralph Baer, Bill Harrison, and Bill Rusch had already built a ping-pong video game for the Brown Box in November 1967 and filed for a patent application on January 15, 1968 (methodical and detail-oriented to a fault, Baer documents everything). Magnavox acquired the license to manufacture the Brown Box on March 3, 1971, and began demonstrating its new home video game console (the Odyssey) to dealers a year later. US Patent 3,659,285 "Television Gaming Apparatus and Method," was awarded to Harrison, Rusch, and Baer on April 25, 1972. In the beginning of May 1972, Magnavox started

bringing the Odyssey to trade shows; later that month, Nolan Bushnell attended an event at the Airport Marina Hotel in Burlingame, CA, where he both saw and played the Odyssey's ping-pong video game. This was a month before the founding of Atari and years after Baer's patent application.

7. "I found out later that this was simply an exercise that Nolan gave me because it was the simplest game that he could think of. He didn't think it had any play value . . . He was just going to dispose of it anyway." Al Alcorn, quoted in Steven L. Kent, *The Ultimate History of Video Games: The Story Behind the Craze that Touched our Lives and Changed the World* (New York: Three Rivers Press, 2001), 41.

8. Steven L. Kent, *The Ultimate History of Video Games: The Story Behind the Craze that Touched our Lives and Changed the World* (New York: Three Rivers Press, 2001), 43.

9. Ralph Baer had a hand in this machine as well. Connecticut-based toy company Coleco—a recent entry into video gaming—had a problem with radio frequency interference related to the Telestar console and was at risk of losing FCC approval. Baer figured out a solution and Coleco was able to implement it in time to receive approval and make it to market by Father's Day in 1976. They sold over $100 million in new consoles and rose to the top—if only briefly—of the home video game market. Steven L. Kent, *The Ultimate History of Video Games: The Story Behind the Craze that Touched our Lives and Changed the World* (New York: Three Rivers Press, 2001), 96–99.

10. Because the Channel F used interchangeable cartridges, the number and variety of games you could play on it was much greater than *Home Pong*, which had just one. Though it never developed much of a following, the Channel F effectively killed *Home Pong* and changed the home console industry forever.

11. The Entertainment Software Association, *2011 Essential Facts About the Computer and Video Game Industry* (The Entertainment Software Association, 2011), 2, accessed September 5, 2012, http://www.theesa.com/facts/pdfs/ESA_EF_2011.

3. LET THE GAMES BEGIN

1. For the full story, I highly recommend watching Picturehouse's *King of Kong: A Fistful of Quarters*, directed by Seth Gordon. However, for those who can't wait to find out what happened . . . Steve Wiebe won. On June 3, 2005, Wiebe set a new *Donkey Kong* world record of 985,600 in front of a crowd of gamers at the Funspot arcade in Weirs Beach, New Hampshire—obliterating Billy Mitchell's August 13, 1982, high score of 874,300. The next day, however,

a game play videotape submitted by Mitchell (and verified by Twin Galaxies, the official scorekeeper for classic arcade games) showed him reaching a score of 1,047,200. This put him back on top, but only temporarily: Wiebe regained the title on August 3, 2006, with a score of 1,049,100. Since then, Wiebe and Mitchell have traded places a few times, but on December 27, 2010, they were both surpassed by Dr. Hank Chien, a New York–based plastic surgeon who posted 1,068,000 points. Chien then went on to break his own record four times. As of July 25, 2012, Chien's *Donkey Kong* world record score stands at 1,127,700. It is sure to be broken.

2. So decreed by then mayor Jerry Parker, and officially recognized by Iowa Governor Terry Branstad, US Senator Charles Grassley, Atari, Inc., and the Amusement Game Manufacturers Association. The designation came about due to Ottumwa's status as the home of Twin Galaxies, official compiler and gatekeeper of video game world records.

3. Sanders, it turned out, was lying. At one point, he had a *Donkey Kong* high score of around 450,000 (still far below Mitchell's world record 874,600), but he'd lost that to another player well before the Ottumwa event. Rather than try to retake it, Sanders simply claimed he'd scored more to put him back on top. The truth came out after he failed to break 200,000 even once during the Video Olympics—not a hard score to achieve for someone who'd purportedly cleared more than twice that. Peter Rugg, "For Disgraced Former Joust King, There's Life After the Arcade," *The Pitch*, January 5, 2009, accessed January 23, 2013, http://www.pitch.com/kansascity/for-disgraced-former-joust-king-steve-sanders-theres-life-after-the-arcade/Content?oid=2193027&showFullText=true.

4. Munoz actually coined the term "cyberathlete." He'd been reading William Gibson's *Neuromancer* at the time, and loved the idea of Gibson's cyberspace—where people could, through use of a computer interface, move freely through a virtual world that linked real people together and was governed by information. Munoz saw a direct connection between Gibson's creation and the pro gamers who were engaging in contests of athletic skill within a virtual arena. He took cyber and athlete and merged them into one descriptive word: cyberathlete.

5. First-person shooters (FPS) are video games where the player sees the world from a first-person perspective: Your view into the virtual environment is exactly as it would be if you were looking out on a real environment. You see your hands in front of you, along with whatever weapon your character is holding. If you look down in the game, you see your feet. Objects shrink with increased distance and get larger as you approach them. In a single-player FPS (in other words, just you against the machine), you're wandering around inside this virtual world fighting computer-controlled enemies who are trying to kill

you and/or prevent you from reaching your objectives. In a multiplayer situation, it's you against enemies controlled by other gamers—you're fighting real people within the virtual world. This is deathmatch, and it's the basis for many of the games played in competition today.

6. "Fragging on the Verge," *Game Informer Magazine*, January 2007, 40–43.

7. That's not a misspelling. His moniker derives from a lexicon known as leet, or leetspeak. "Leet" is short for "elite" and has its origins in 1980s computer bulletin board systems, where elite status granted users access to generally restricted files, folders, or online chat rooms. It's since been adopted by Internet and gaming culture, particularly with respect to creating player aliases (known as gamertags). At its most basic, leet words are formed by substituting numbers and/or symbols for similarly shaped letters. For example, leet would be rendered "1337." In Wendell's case, he simply replaced the letter I with the number 1—but he could easily have gone farther, to F474l17y, for instance (the numbers 4 and 7 replacing the letters A and T, respectively). Other leet words come from typos or alternate, sometimes phonetic, spellings. Two of the most common are pwn and n00b. Pwn translates to "own" (a frequent typo, as O and P are neighbors on a standard computer keyboard), which loosely means "to beat somebody severely." N00b is an abbreviation for "newbie" (the shortened form "newb" recast as "noob," and then zero substituted for the letter O)—a term used to describe anyone new to, in this case, gaming. For more on leet, Wikipedia is a good starting point ("Leet," last modified January 2013, Simple English Wikipedia, http://simple.wikipedia.org/wiki/Leet). If you're curious enough to try it out for yourself, you can also find a leet translator online at http://www.albinoblacksheep.com/text/leet.

8. Like many of the best cyberathletes, Fatal1ty is in excellent physical condition. He works out regularly, running a couple of miles a day, particularly when preparing for a big tournament. This may seem strange to the uninitiated—that physical fitness would be crucial to a relatively unphysical sport—but some competitions can last three days, with not a lot of opportunity for rest. Performing at the level required to just place, never mind win, takes stamina and mental acuity. For gaming pros at Fatal1ty's level, fitness is critical. In his experience, "It's all about who has the most stamina and can think faster." Jeremy Caplan and Ta-Nehisi Paul Coates, "Tiger. Jordan. Hawk. Wendel?," *TIME*, February 2007, 60–61.

9. Though many of the circumstances surrounding the Cyberathlete Professional League's (CPL's) sale are clouded by controversy, and opinions about Munoz himself run the gamut from vitriolic demonification to wholehearted support, the reality, according to Munoz, is simple. The league peaked in 2004, and then in 2005, it launched a globe-spanning tour with 10 stops around the

world. "That single tour strained the entire CPL staff to the point of complete exhaustion," he said. "Staff attrition was destabilizing our operations and I somewhat realized that we had hit a ceiling for what could be accomplished by the CPL in professional gaming." Combined with the rise of several competing leagues, the looming global recession, and the failure of key sponsors to follow through on their financial commitments (with respect to tournament prizes), Munoz felt that, for the CPL and for him personally, the writing was on the wall. Andrei Hancu, "Angel Munoz: 'The Apex was 2004,'" *HTLV.org*, October 8, 2010, accessed February 3, 2013, http://www.hltv.org/news/5443-angel-munoz-the-apex-was-2004&nc=1.

10. "Fragging on the Verge," *Game Informer Magazine*, January 2007, 40–43.

11. World Cyber Games, WCG Concept, accessed February 3, 2013, http://www.wcg.com/renew/inside/wcgc/wcgc_concept.asp.

12. "Fragging on the Verge," *Game Informer Magazine*, January 2007, 40–43.

13. Eric A. Taub, "Taking Their Game to the Next Level," *New York Times*, October 7, 2004, Technology, accessed February 3, 2013, http://www.nytimes.com/2004/10/07/technology/circuits/07play.html?pagewanted=1&_r=0.

14. Richard Nieva, "Video Gaming on the Pro Tour, for Glory but Little Gold," *New York Times*, November 28, 2012, Personal Tech, accessed February 3, 2013, http://www.nytimes.com/2012/11/29/technology/personaltech/video-gaming-on-the-pro-tour-for-glory-but-little-gold.html?_r=0.

15. "Fragging on the Verge," *Game Informer Magazine*, January 2007, 40–43.

16. Blood Gulch is the most famous multiplayer level from *Halo: Combat Evolved*.

17. They behaved like fans at any other sporting event: cheering the home favorites, booing and heckling their opponents. And they appreciated, with much vocal enthusiasm, the regular displays of skill: the jumping headshot, the killing streaks, the clutch recovery of the flag mere inches from the opposing team's base. They'd chant a particular player's name out of respect for an inspiring performance, to encourage a down favorite, or to acknowledge a particular achievement. And they'd groan when one of their favorites was hit.

18. Jon Robinson, "Major League Gaming continues to grow," *ESPN.com*, August 24, 2012, accessed February 3, 2013, http://espn.go.com/blog/playbook/tech/post/_/id/1900/major-league-gaming-continues-to-grow.

19. Alex Wilhelm, "How Major League Gaming went All-in, Landed on Its Feet, and Raised Millions," *TNW: The Next Web*, March 13, 2012, accessed February 3, 2013, http://thenextweb.com/insider/2012/03/13/how-major-league-gaming-went-all-in-landed-on-its-feet-and-raised-millions/.

20. Kim Bhasin, "How Sundance DiGiovanni Built The Largest Competitive Gaming League In America," *Business Insider*, May 1, 2012, accessed February 3, 2013, http://articles.businessinsider.com/2012-05-01/strategy/31515986_1_esports-video-game-major-league-gaming.

21. Paul Tassi, "Talking the Past, Present and Future of eSports with MLG's Sundance DiGiovanni," *Forbes*, August 22, 2012, accessed February 3, 2013, http://www.forbes.com/sites/insertcoin/2012/08/22/talking-the-past-present-and-future-of-esports-with-mlgs-sundance-digiovanni/.

22. Jon Robinson, "Major League Gaming continues to grow," *ESPN.com*, August 24, 2012, accessed February 3, 2013, http://espn.go.com/blog/playbook/tech/post/_/id/1900/major-league-gaming-continues-to-grow.

23. Chelsea Stark, "Move Over, Super Bowl. Spectator Gaming Reaches Millions Online," *Mashable.com*, February 27, 2012, accessed February 3, 2013, http://mashable.com/2012/02/27/spectator-gaming-esports/.

24. Kim Bhasin, "How Sundance DiGiovanni Built The Largest Competitive Gaming League In America," *Business Insider*, May 1, 2012, accessed February 3, 2013, http://articles.businessinsider.com/2012-05-01/strategy/31515986_1_esports-video-game-major-league-gaming.

25. *Ibid.*

4. ALPHABET SOUP

1. A persistent-world game is an online game whose world continues to exist and develop even when players are offline. For example, if you happened to be playing *EVE* and you ended your session, events within the world would continue—and other players could keep playing—while you were away. This is unlike playing, say, *SimCity* on your Mac or PC, where the game stops progressing when you quit and starts up again when you return. A massively multiplayer online role-playing game is a game played online in which you create a character (your role) and can interact with potentially a very large number of other players within the game's virtual world.

2. The Ubiqua Seraph (UQS) incident is perhaps the best example of the kind of treachery *EVE* allows players to conduct. Over a period of a year, Guiding Hand Social Club (GHSC) agents worked their way into stations of power and trust within UQS—including the number two spot in Mirial's organization—putting themselves in a position to inflict maximum damage and financial disruption. When the fateful day arrived, the GHSC executed a near-perfect strike, killing Mirial and many of her top officers, destroying ships and buildings, stealing original blueprints (which are required to build starships in *EVE*), and emptying the coffers of 30 billion ISK (*EVE*'s game currency). The

destruction and/or theft of materials and documents amounted to another 20 billion ISK, which, combined with the 30 billion heist, equated to a total real-world value of about $16,500. Building UQS to the level of power it held just prior to the GHSC attack also took Mirial about three years, so she lost not only money but time (and skill points accrued over that time period). To call this a devastating loss is a gross understatement: UQS collapsed and Mirial, to everyone's knowledge, hasn't been seen since. You can read the full story in *PC Gamer* magazine. Tom Francis, "Murder Incorporated: The Story of *EVE Online*'s Most Devastating Assassins," *PC Gamer Magazine*, January 2006, 90-106, archived at *ComputerAndVideoGames.com*, January 29, 2008, accessed February 16, 2013. http://www.computerandvideogames.com/180867/features/murder-incorporated/.

3. Ethic, "Interview: Istvaan Shogaatsu," *Kill Ten Rats*, October 12, 2005, accessed February 16, 2013. http://www.killtenrats.com/2005/10/12/interview-istvaan-shogaatsu/.

4. Stewart Alsop II, "TSR Hobbies Mixes Fact and Fantasy," *Inc.*, February 1, 1982, accessed February 8, 2013. http://www.inc.com/magazine/19820201/3601.html.

5. *Ibid.*

6. Interactive fiction games are text-based adventures that present users with a complete narrative that unfolds according to their actions. The story is, of course, prescripted, but because it's revealed in pieces and seems to be affected by the choices you make, it has the appearance of interactivity. Without a doubt, you *are* interacting with the game, but only insofar as you're deciding what route to take down a predetermined path.

7. *Adventure*, in turn, grew out of an even earlier game: William Crowther's *Colossal Cave*. Crowther created the game in 1976 as a way to maintain contact with his daughters after he and his wife divorced. (Brad King and John Borland, *Dungeons & Dreamers: The Rise of Computer Game Culture From Geek to Chic*, [Blacklick, Ohio: McGraw-Hill Osborne Media, 2003], 31). An avid caver, he drew on Kentucky's Mammoth Cave system for part of the game's layout. Don Woods discovered *Colossal Cave* in 1977 while undertaking graduate studies at Stanford University. With Crowther's blessing, Woods cleaned up and expanded the game, adding many fantasy elements based on the writings of J.R.R. Tolkein, and calling the finished game *Adventure*. Though other programmers have modified *Colossal Cave* over the years, Woods' version is the best-known variant. *Zork*—developed at Massachusetts Institute of Technology by Marc Blank, Tim Andersen, Bruce Daniels, and Dave Liebling—took *Adventure* to another level, adding even more elements and greatly enlarging the game world.

8. Brad King and John Borland, *Dungeons & Dreamers: The Rise of Computer Game Culture From Geek to Chic* (Blacklick, Ohio: McGraw-Hill Osborne Media, 2003), 54.

9. Garriott's friends gave him the moniker British as a teenager, due to his habit of saying "hello" when answering the phone, rather than the shorter, informal, and distinctly American "hi." Lord was added as a result of his role as Dungeon Master during games of *Dungeons & Dragons*.

10. Shortly after its 1997 release, *Ultima Online* became the first massively multiplayer online role-playing game to reach 100,000 subscribers. In 2003, it peaked at close to 250,000 before sliding back to around 100,000 in 2008. Electronic Arts, *EA Announces Ultima Online (TM): Kingdom Reborn (Working Title); The Game That Firmly Established the MMORPG Genre Receives a Massive Visual Overhaul and New Content in 2007* (Electronic Arts, 2006), accessed February 10, 2013, http://investor.ea.com/releasedetail.cfm?ReleaseID=314331.

11. For almost all massively multiplayer online role-playing games (MMOs), the game doesn't run on one machine but is distributed across several game servers. This is how *World of Warcraft* operates, and so it *is* possible for two players to have the same rare, one-of-a-kind item—if they are playing on two different servers. Those players would never meet up in the game, however, because each time they start up *WoW* on their home computers, they're directed to two separate servers to play. This means that if all 9.6 million *WoW* players were somehow to play the game simultaneously, they wouldn't all encounter one another; only those on the same servers would have the opportunity to meet. It's useful to think of each server as its own distinct world. *World of Warcraft*, then, exists as multiple worlds. *EVE Online* is unique among MMOs in that there is only a single game server, and every user plays on it. *EVE* is one massive world, which means that, theoretically, every player could interact with every other player at some point. Highly unlikely, but still possible.

12. Bear in mind, that 190 hours is for a *very* proficient gamer. For the average person, it could easily be double that—or around 10 work weeks.

13. Like trading US dollars for Euros, game currencies are subject to fluctuating exchange rates against real-world currencies. On February 13, 2013, the exchange rate between USD and ISK was approximately $10 to ISK 460 million. For Euros, it was slightly better: 7.4 € to ISK 460 million.

14. Gold farming encompasses more than just the sale of virtual currency. There are special items within MMOs—weapons that inflict extra damage, armor that affords additional protection, rare spell books, etc.— that occur infrequently, and for which gamers in need will pay a premium. Scarce and/or

finite resources, such as raw materials needed to concoct potions, also find a steady market of buyers.

15. Dibbell chronicles his 12-month exploit in *Play Money*. One of the more depressing aspects he reveals is that, though as a gamer he truly enjoyed playing *Ultima Online*, the experience of transforming it into work effectively killed anything fun about the game, and it just became a daily grind to try and make a living—the danger of turning any hobby into a profession.

16. Julian Dibbell, *Play Money, Or, How I Quit My Day Job and Made Millions Trading Virtual Loot* (New York: Basic Books, 2006), 302–313.

17. Julian Dibbell, "The Kingpin of Azeroth," *Wired*, December 2008, 216.

18. Julian Dibbell, *Play Money*, 12.

19. Julian Dibbell, "The Kingpin of Azeroth," *Wired*, December 2008, 182.

20. Mark Anderson, "The End of Gold Farming?" *IEEE Spectrum*, October 2010, accessed February 16, 2013. http://spectrum.ieee.org/consumer-electronics/gaming/the-end-of-gold-farming.

21. Newzoo, *The Global MMO Market; Sizing and Seizing Opportunities* (NewZoo, 2012), accessed February 16, 2013. http://www.newzoo.com/infographics/the-global-mmo-market-sizing-and-seizing-opportunities/.

22. *EVE*'s developer, CCP, is based in Iceland. As of 2013, Iceland's population was 300,000. *EVE*'s was 400,000 and growing.

23. Internet Game Exchange (www.igxe.com) is one of the most prominent of the recent entries.

24. According to Alex Engel, this is the modern way to gold farm. "Now . . . they're hacking players' accounts and using hacked accounts to funnel money out of them to the buyers. Say a buyer says, 'I want to buy 200 gold.' They'll just hack however many accounts they need to reach that 200 gold threshold."

5. NO CONSOLE REQUIRED

1. Newzoo, *Online Casual & Social Games Trend Report* (Newzoo, 2012), 3, accessed December 21, 2012. http://www.newzoo.com/trend-reports/casual-social-games-trend-report/.

2. Matt Hulett, personal communication.

3. Information Solutions Group, *2011 PopCap Games Social Gaming Research* (Information Solutions Group, 2012), 22, accessed December 21, 2012. http://www.infosolutionsgroup.com/pdfs/2011_PopCap_Social_Gaming_Research_Results.pdf.

4. eMarketer. *Mobile, Social Boost Online Gaming Populations* (eMarketer, 2012), accessed December 21, 2012. http://www.emarketer.com/Article.aspx?R=1009100.

5. Game company jobs and titles are extremely fluid, and people move around with dizzying rapidity. Between my interview with Matt Hulett in Seattle and the publication of this book, he'd left GameHouse.

6. Casual Games Association, *Social Network Games 2012, Casual Games Sector Report* (Casual Games Association, 2012), accessed December 21, 2012. http://dl.dropbox.com/u/3698805/research/2012_CGA_SocialSector.pdf.

7. Newzoo, *Mobile Games Trend Report* (Newzoo, 2012), 4, accessed December 21, 2012. http://www.newzoo.com/trend-reports/mobile-games-trend-report/.

8. Seth Schiesel has a great review of *Shadow Cities* in *The New York Times.* Seth Schiesel, "Brave New World That's as Familiar As the Machine It Fights With," *The New York Times*, July 6, 2011, The Arts, C1.

9. Casual Games Association, *Mobile Gaming 2012 Sector Report* (Casual Games Association, 2012), accessed December 21, 2012. http://dl.dropbox.com/u/3698805/research/2012_CGA_MobileSector.pdf.

10. Between the time we spoke and the publication of this book, Peter Hofstede left Spil Games and went on to found his own game engineering company, Wungi.

11. Information Solutions Group, *2011 PopCap Games Social Gaming Research* (Information Solutions Group, 2012), 22, accessed December 21, 2012. http://www.infosolutionsgroup.com/pdfs/2011_PopCap_Social_Gaming_Research_Results.pdf.

12. The iPad exemplifies the speed of technological progress. It took more than 10 times longer—nearly 34 years—to go from Martin Cooper's first handheld phone call (April 3, 1973, for those keeping score at home) to the first iPhone, which debuted on January 9, 2007. By contrast, Steve Jobs revealed the first iPad on January 27, 2010.

13. Arcuri's since left Zipline Games and is now with the Kirkland, Washington–based INRIX.

14. Andrew Ross Sorkin, Editor-At-Large, "Angry Birds Maker Posted Revenue of $106.3 Million in 2011," *New York Times Dealbook*, May 7, 2012, accessed December 21, 2012. http://dealbook.nytimes.com/2012/05/07/angry-birds-maker-posts-2011-revenue-of-106-3-million/.

15. *Ibid.*

16. Newzoo, *Online Casual & Social Games Trend Report* (Newzoo, 2012), 3, accessed December 21, 2012. http://www.newzoo.com/trend-reports/casual-social-games-trend-report/.

17. *Ibid*, 2.

18. Information Solutions Group, *2011 PopCap Games Social Gaming Research* (Information Solutions Group, 2012), 23, accessed December 21, 2012.

http://www.infosolutionsgroup.com/pdfs/2011_PopCap_Social_Gaming_Research_Results.pdf.

19. Casual Games Association, *Social Network Games 2012, Casual Games Sector Report* (Casual Games Association, 2012), accessed December 21, 2012. http://dl.dropbox.com/u/3698805/research/2012_CGA_SocialSector.pdf; Casual Games Association, *Mobile Gaming 2012 Sector Report* (Casual Games Association, 2012), accessed December 21, 2012. http://dl.dropbox.com/u/3698805/research/2012_CGA_MobileSector.pdf.

20. IGDA, *2008–2009 Casual Games Whitepaper* (International Game Developers Association, 2009), 171, accessed December 21, 2012. http://www.igda.org/sites/default/files/IGDA_Casual_Games_White_Paper_2008.pdf.

6. DRESSED FOR THE SYMPHONY

1. When composing musical themes for *God of War*, Gerard Marino tried to capture the main character's rage, desire for revenge, and need, ultimately, for redemption. As the game was based on Greek mythology, he based many of the musical rhythms on the Greek words for revenge and redemption. For the history-spanning strategy game *Civilization IV*, composer Christopher Tin fused classical and world music to reflect the game's temporal sweep and create its signature sound.

2. Interestingly, what they're hearing in the live orchestra is its imperfection. In Brick's experience, "synthesizers and samplers are too perfect, and the human ear and psyche doesn't react well to that. The imperfection in the orchestra . . . it's the fact that every string player, even if they're playing the same note, is playing slightly out of tune from the guy or girl next to them. The fact that 40 string players are all playing slightly out of tune gives it a richness. When you have a computer and synthesizer, it plays exactly in tune. And we don't like that."

3. For those who don't, though, it's been anything but. A composer who relies too heavily on computers and samplers faces the very real danger of being out of a job if a game developer wants to go with a live orchestra. As Brick told me, "I can't tell you how many calls I've gotten from younger guys who have been hired to write music for a game, the developer liking what they've done in their demos with the synthesizers and computers, then telling them, 'Okay, we're gonna go with a live orchestra. Here's $100,000. Go record it.' And they just kind of go, 'What? How do I do that?' A lot of guys wind up reaching the end of their careers at that moment."

4. Asian markets have historically been much more open and accepting of video games as a part of their culture, so a string of successful video game

music concerts held in Japan is not surprising. Europe and North America—the other major video game markets—have always lagged behind Asia with respect to their outlook on games. That the Symphonic Game Music Concert series took place in Europe (albeit a dozen years after the first Japan concert) is, therefore, far more notable. Not being Euro-centric here, just telling it like it is.

5. *Video Games Live Level 2*, directed by Allen Newman (Sherman Oaks, CA: Mystical Stone Entertainment, 2010), DVD.

6. Video Games Live is especially popular in South America. For the last seven years, they've sold out every show on the continent. "We're huge down in Brazil," says Tallarico, "and the government helps to subsidize the whole thing because they see the value in getting young people interested in the arts and culture." Sharon Eberson, "PSO presents 'Video Games Live' to turn on new generation," *Pittsburgh Post-Gazette*, July 7, 2009, accessed December 6, 2012, http://www.post-gazette.com/stories/sectionfront/life/pso-presents-video-games-live-to-turn-on-new-generation-348458/.

7. *Ibid.*

8. *Ibid.*

7. FROM THE FLAT SCREEN TO THE BIG SCREEN

1. Clive Thompson, "The Science of Play," *Wired Magazine*, September 2007, 144. Thumbstick refers to the standard Xbox game controller, which, in addition to a dizzying array of buttons, features two miniature joysticks controlled by dexterous and exacting movements of a gamer's thumbs.

2. Though it's fairly dry reading, Wikipedia has a list of the various media adaptations of the series in its online encyclopedia. "List of *Halo* Media," last modified January 2013, Wikipedia, The Free Encyclopedia, http://en.wikipedia.org/wiki/List_of_Halo_media.

3. Next generation consoles refer to game consoles available at the turn of the twenty-first century, such as the Sega Dreamcast, Sony PlayStation 2, Nintendo Gamecube, and Microsoft Xbox (the only game system *Halo* ever appeared on—though it was eventually released for both the Mac and Windows operating systems). Microsoft, *Halo: Combat Evolved for Xbox Tops 1 Million Mark In Record Time* (Microsoft, 2002), accessed January 15, 2013, http://www.microsoft.com/en-us/news/press/2002/apr02/04-08halomillionpr.aspx.

4. Wikipedia's entry on *Halo: Combat Evolved* provides a good summary of the game's awards and achievements. "*Halo: Combat Evolved*," last mod-

ified January 2013, Wikipedia, The Free Encyclopedia, http://en.wikipedia.org/wiki/Halo:_Combat_Evolved#Reception.

5. Michael Marriott, "Hollywood Would Kill for Those Numbers," *New York Times*, November 14, 2004 Week in Review, accessed January 15, 2013, http://www.nytimes.com/2004/11/14/weekinreview/14bigp.html?_r=0.

6. Though Microsoft owns *Halo*, the developers at Bungie are the creative geniuses behind the game. As is the way of many things, their success with prior games—especially the *Marathon* trilogy—and the preliminary demo footage for *Halo* brought them to Microsoft's attention, which promptly bought the formerly independent game developer and brought them in-house as part of Microsoft Game Studios.

7. As of January 2013, the fastest movie to reach this benchmark is *The Avengers*, which hit $150 million in two days—still less than *Halo 3*'s domestic one-day sales. "All Time Box Office," Box Office Mojo, http://www.boxofficemojo.com/alltime/.

8. *Ibid.*

9. Open-world games feature expansive virtual environments that allow players to explore as they wish. With such games, you can spend as much (or more) time exploring the world as you can battling enemies or completing quests.

10. Robert A. Lehrman, "I Play Therefore I Am," *Christian Science Monitor*, March 2012, 26–31. Granted, this represents global sales, but it's an impressive figure nonetheless.

11. "All Time Box Office," Box Office Mojo, http://www.boxofficemojo.com/alltime/.

12. Marvel, *Avengers Shatters More Records* (Marvel, 2012), accessed January 15, 2013, http://marvel.com/news/story/18690/avengers_shatters_more_records.

13. Activision, *Call Of Duty®: Black Ops II Delivers More Than $500 Million In Worldwide Retail Sales In First 24 Hours* (Activision, 2012), accessed January 15, 2013, http://investor.activision.com/releasedetail.cfm?ReleaseID=721903.

It smashed the $400 million first-day record held only a year before by the previous entry in the series—*Modern Warfare III*.

14. Activision, *Call of Duty®: Black Ops II Grosses $1 Billion In 15 Days* (Activision, 2012), accessed January 15, 2013, http://investor.activision.com/releasedetail.cfm?ReleaseID=725026.

15. Motion Picture Association of America, *2011 Theatrical Market Statistics* (Motion Picture Association of America, 2012), 2, accessed September 5, 2012, http://www.mpaa.org/resources/5bec4ac9-a95e-443b-987b-bff6fb5455a9.pdf.

16. The Entertainment Software Association, *2012 Essential Facts About the Computer and Video Game Industry* (The Entertainment Software Association, 2012), 11, accessed September 5, 2012, http://www.theesa.com/facts/pdfs/ESA_EF_2012.pdf.

17. Motion Picture Association of America, *2011 Theatrical Market Statistics* (Motion Picture Association of America, 2012), 2, accessed September 5, 2012, http://www.mpaa.org/resources/5bec4ac9-a95e-443b-987b-bff6fb5455a9.pdf.

18. Peter Warman, "The Bigger Picture: Key Facts and Trends on the Ever-changing Global Games Market," NewZoo (keynote delivered at LU-CIX conference, Luxembourg, September 2012), 8, accessed January 15, 2013, http://www.newzoo.com/wp-content/uploads/Luxembourg_gamingevent-Newzoo_Keynote.pdf.

19. [illegible], "A [illegible] That Takes Aim at Bigger Screens," *New York Times*, November 7, 2009, The Arts, C1.

20. *Ibid.*

21. *Ibid.*

22. A feat that took him several hours longer than expected because at times he was so distracted by what he was seeing that he forgot to play and kept dying. "The very first scene in *Uncharted* 2 came on the screen," he wrote, "and I was literally stunned into a moment of cognitive dissonance. Just for a second, what I was seeing could have passed for a live-action shot from a mountainside in Tibet. For a moment I thought, what am I looking at?" *Ibid.*

23. Robert Alan Brookey, *Hollywood Gamers: Digital Convergence in the Film and Video Game Industries* (Bloomington: Indiana University Press, 2010), 3.

24. Emru Townsend, "The 10 Worst Games of All Time," *PCWorld*, October 2006, 2, http://www.pcworld.com/article/127579/article.html?page=2.

25. Steven L. Kent, *The Ultimate History of Video Games: The Story Behind the Craze that Touched our Lives and Changed the World* (New York: Three Rivers Press, 2001), 237–240.

26. Wikipedia has a good discussion of the game, as well as links to articles related to its effects on Atari and the video game industry as a whole. "*E.T. the Extra-Terrestrial* (video game)," last modified January 2013, Wikipedia, The Free Encyclopedia, http://en.wikipedia.org/wiki/E.T._the_Extra-Terrestrial_(video_game).

27. Michael Wilmington, "No Offense Nintendo: Super Mario Bros. Jump to Big Screen in Feeble Extravaganza," *Los Angeles Times*, May 29, 1993, accessed January 10, 2013, http://articles.latimes.com/1993-05-29/entertainment/ca-41093_1_super-mario-bros.

28. Simon Hattenstone, "The Method? Living it out? Cobblers!," *The Guardian*, August 2, 2007, accessed January 10, 2013, http://www.guardian.co.uk/film/2007/aug/03/2.

29. Robert Alan Brookey, *Hollywood Gamers: Digital Convergence in the Film and Video Game Industries* (Bloomington: Indiana University Press, 2010), 4.

30. The most successful video game–based film series thus far, the *Resident Evil* franchise has brought in more than $870 million worldwide on a budget of about $280 million. Box Office Mojo has all the details. "Resident Evil," Box Office Mojo, http://www.boxofficemojo.com/search/?q=Resident%20Evil.

31. Anne Thompson, "What Will Tron: Legacy's 3D VFX Look Like in 30 Years?" *Popular Mechanics*, December 2010, http://www.popularmechanics.com/technology/digital/visual-effects/are-tron-legacys-3d-fx-ahead-of-their-time.

32. David Thier, "Summer 2010: When Video Games Took Over the Movies," *The Atlantic*, August 2010, http://www.theatlantic.com/entertainment/archive/2010/08/summer-2010-when-video-games-took-over-the-movies/61522/.

33. Polygons are simply basic shapes used as building blocks for digital models. Rendering refers to the process of taking a rough digital model and preparing (smoothing) it for use in the finished movie or video game.

34. Shane Culp, personal communication.

35. Sadly, very few people will ever get to experience this world. In October of 2012, Disney purchased Lucasfilm for four billion dollars . . . and we all held our collective breath. Then, on April 3, 2013, the other shoe dropped: Disney announced that it was shutting down LucasArts and canceling all further game development—including *Star Wars 1313*. However, this doesn't change the fact of what I saw in the demo room, nor does it negate the groundbreaking achievement of ILM and LucasArts. The technology, ability, and talent to produce a game like *1313* are still out there, and so it's inevitable that we, as gamers, will be treated to an immersive, mature, and fully realized photorealistic game world at some point. *Star Wars* just won't be a part of it.

36. As noted in 1975 by Intel executive David House, one result of Moore's Law was that overall computer performance would double every eighteen months. In actuality, it does so in around twenty months, but he was close.

37. That million-fold increase in speed over a decade actually considerably outstrips Moore's Law, which would predict an increase over the same time period of around 16 times.

8. VIRTUAL LIFE

1. The word *avatar* has its roots in Hindu mythology, in which it refers to the physical manifestation of a god—usually Vishnu.

2. Many people refer to *Second Life* and other virtual worlds as games. I'm hesitant to follow this convention. While virtual worlds and games do share some common traits (they're essentially massively multiplayer online games), virtual worlds have no defined end-state—there's no way to win, except possibly to succeed at meeting goals that users set for themselves (and then that's only winning on an individual level). Virtual worlds move beyond objective-based gaming to a hybrid of gaming and real-world interaction, which can give them broader social impact. As such, they can't *really* be called games (not if we're being honest, anyway).

3. Julian Dibbell, "Griefer Madness," *Wired*, February 2008, 90–97.

4. They divorced on the grounds of "unreasonable behavior."

5. Steven Morris, "Second Life Affair Leads to Real Life Divorce," *The Guardian*, November 13, 2008, accessed February 19, 2013, http://www.guardian.co.uk/technology/2008/nov/13/second-life-divorce.

6. Qiu Chengwei killed fellow *Legends of Mir 3* player Zhu Caoyuan after he had sold a powerful Dragon Sabre loaned to him by Chengwei. Chengwei complained to the police about the theft, but they declined to help, stating that the weapon was "not real property." Seeing no other recourse, Chengwei tragically took matters into his own hands. "'Game Theft' Led to Fatal Attack," *BBC News*, March 31, 2005, accessed February 19, 2013, http://news.bbc.co.uk/2/hi/technology/4397159.stm.

7. The teenager apparently convinced several players to give him access to their accounts, then stole the furniture, which he and five friends used to decorate their own *Habbo Hotel* rooms. This was considered a punishable crime because you could attach real-world value to the stolen goods: $6,000 was the actual cost of the virtual items. "'Virtual Theft' Leads to Arrest," *BBC News*, November 14, 2007, accessed February 19, 2013, http://news.bbc.co.uk/2/hi/technology/7094764.stm.

8. Josh Katz, "Virtual 'Maple Story' Murder Reveals Online Lives Gone Too Far," *Finding Dulcinea*, October 24, 2008, accessed February 19, 2013, http://www.findingdulcinea.com/news/technology/September-October-08/Virtual--Maple-Story--Murder-Reveals-Online-Lives-Gone-Too-Far.html.

9. Raphael G. Satter, "Virtual Affair Leads Real Divorce For UK Couple," *Huffington Post*, November 2008, accessed February 19, 2013, http://www.huffingtonpost.com/2008/11/14/virtual-affair-yields-rea_n_143860.html.

10. Much like a real-world cultural anthropologist, Boellstorff spent two years conducting fieldwork within *Second Life*, creating an avatar, establishing a home in-world, and living and interacting with its residents.

11. Tom Boellstorff, *Coming Of Age In Second Life* (Princeton: Princeton University Press, 2008), 173.

12. *Ibid*, 170.

13. *Ibid.*

14. *Ibid*, 173.

15. *Ibid*, 170.

16. *Ibid*, 159.

17. This is true of all in-world conversation, regardless of the virtual world. No one speaks of "your avatar" or "your character." They just say "you." It may only be a matter of linguistic convenience: repeatedly saying (or typing, depending on user interface with the virtual world) "your character" would get very cumbersome very quickly. I think that's too simple an explanation, though. This is a phenomenon that goes back at least to the die-and-paper role-playing days of *Dungeons & Dragons*, if not further. When we got together for a *D&D* campaign, we never referred to our characters either. If I was going to attack an Orc, I'd say, "I'm attacking the Orc." There was an immediate and natural transference of my identity into that of my character—at least for the duration of the game. That seems to be what's happening in the context of avatars and virtual worlds today.

18. The Proteus Effect is named for the Greek sea god Proteus, who had the ability to assume many different forms. He also gives his name to the word *protean*—changeable.

19. Nick Yee and Jeremy Bailenson, "The Proteus Effect: The Effect of Transformed Self-Representation on Behavior," *Human Communication Research*, July 2007, 271–290.

20. Nick Yee, Jeremy N. Bailenson, and Nicolas Ducheneaut, "The Proteus Effect: Implications of Transformed Digital Self-Representation on Online and Offline Behavior," *Communication Research*, April 2009, 285–312.

21. Maria Korolov, personal communication.

22. The MOO in *LambdaMOO* stands for MUD (multiuser dungeon), Object-Oriented. It's a text-only virtual world that stores every individual item in the world (chairs, lamps, rugs, trees, rocks, etc.) as a discrete object in a database. Each location or environment in the world is filled by programming specific database objects to appear in specific locations—for example, rugs, lamps, and chairs in a sitting room or trees and rocks in a forest. Visitors can wander through the world at will or, if they're adept at it, can build new sections of it through object-oriented programming—essentially asking certain items contained within the database to appear in certain places.

23. Legba is one of the primary Haitian spirits, the intermediary for communication between other Haitian spirits and God.

24. Julian Dibbell, "A Rape in Cyberspace," juliandibbell.com. Originally published in *The Village Voice*, December 23, 1993. Accessed March 20, 2013, http://www.juliandibbell.com/articles/a-rape-in-cyberspace/.

25. *Ibid.*

26. Jesse Fox, Jeremy N. Bailenson, and Liz Tricase, "The Embodiment of Sexualized Virtual Selves: The Proteus Effect and Experiences of Self-Objectification via Avatars," *Computers in Human Behavior*, 2013, 930–938.

27. Diane M. Quinn, Rachel W. Kallen, and Christie Cathey, "Body on My Mind: The Lingering Effect of State Self-Objectification," *Sex Roles*, December 2006, 869–874.

28. Jesse Fox, Jeremy N. Bailenson, and Liz Tricase, "The Embodiment of Sexualized Virtual Selves: The Proteus Effect and Experiences of Self-Objectification via Avatars," *Computers in Human Behavior*, 2013, 930–938.

29. Anil Ananthaswamy, "A Life Less Ordinary Offers Far More Than Just Escapism," *New Scientist*, August 25, 2007, 26–27.

30. *Ibid.*

9. . . . AND WE ARE MERELY PLAYERS

1. Blizzard did this by giving nonplayer characters an ungodly amount of hit points. Hit points (a system borrowed from *Dungeons & Dragons*) indicate the amount of damage you can take, represented by a number. Characters start out with relatively few hit points, which increase as they gain experience and level up. When you take damage, hit points drop; when they reach zero, you're dead. Players who have healing abilities can restore hit points and even resurrect dead characters.

2. Fefferman Lab. "Nina H. Fefferman, PhD." Accessed March 7, 2013, http://rci.rutgers.edu/~feffermn/fefferman.php.

3. She also laughs a lot, which may be her way of staying sane while studying organisms that could end human civilization as we know it.

4. There's something about *World of Warcraft* that's important to understand: While the general perception is that video games skew to the young male demographic (and many do), *World of Warcraft*, at least during the time of Corrupted Blood, had a uniquely diverse population that represented all strata of society. There were students, doctors, working moms, deployed soldiers, professionals, academics . . . and, yes, 18- to 24-year-old males.

5. Fire spread is a good example of this. When firefighters are trying to contain a blaze, they need to think about where to direct water to either put a

fire out or prevent one from growing. As Nina explained, "It's very easy, as a human, to say where the fire should go next in the absence of wind, right? And if you forget about wind, your model's still gonna be completely wrong. The lack of having just stuck that thing [wind] in in a very natural way would invalidate the outcome."

6. Although, people do this all the time. After one particular radio interview where a panel member made this same objection, a fire chief emailed her and relayed his own experience. "Part of the biggest problem we have with fire control is keeping stupid people who like fire far enough away from the fire that they don't get burned when the wind changes direction." And, she cautions, it's never wise to underestimate people's stupidity. "Just the sheer number of stupid people who try to break into the White House on frat stunts every year is staggering. Everyone knows it's a terrible idea to try and get past the Secret Service, and yet . . . So, there's boundless stupidity also."

7. However, as she notes, if it occurs to someone to try something in the virtual world—stupid though it may be—it's a decent bet that someone's thought about trying it in the real world, regardless of the possibility of injury or death. "I still don't understand how people are willing to drive without seatbelts but are scared to fly," Nina said. "Or willing to text message while driving. Just because there's a physical risk of real-world death doesn't mean people reject the risk."

8. In Nina's words, it's a process of "mining [the virtual world] for inspiration about the sorts of things we should consider in the real world."

9. A patient zero is typically the first person infected with a disease who, either willfully or accidentally, begins to spread that disease throughout the general population. Typhoid Mary—the New York cook and carrier of typhoid fever—is the most famous case. She transmitted the disease to a large but unknown number of people—many of whom died—but never developed the disease herself.

10. David Thier, "*World of Warcraft* Shines Light on Terror Tactics," *Wired*, March 2008, accessed March 7, 2013, http://www.wired.com/gaming/virtualworlds/news/2008/03/wow_terror.

11. *Ibid.*

12. Yasmin B. Kafai, David Feldon, Deborah Fields, Michael Giang, and Maria Quintero, "Life in the Times of Whypox: A Virtual Epidemic as a Community Event," in *Communities and Technologies*, edited by C. Steinfield, B. Pentland, M. Ackerman, and N. Contractor (New York: Springer, 2007), 171–190. And at the risk of beating a particular dead horse again, let me just add: Proteus Effect.

13. Holli Seitz, "Whyville and the 2009- 2010 Whyflu: Evolving a Virtual World Activity to Meet Changing 'Real World' Communication Needs" (Ab-

stract), *Convergence 2010: National Conference on Health, Communication, Marketing, and Media*, accessed March 7, 2013, https://cdc.confex.com/cdc/nphic10/webprogram/Paper24781.html.

14. The Comanche games developed by Novalogic had controls that were very similar to what you'd find in the Apache games or in the actual Apache helicopter.

15. Lori Mezoff, "Army Game's Medic Training Helps Save Two Lives," *Armyreal.com*, January 22, 2008, http://www.armyreal.com/articles/item/3497.

16. Steven Cattrell, Producer, Virtual Heroes Division. Personal communication.

17. Steve Leblanc, "Studies: Video Games can Make Better Students, Surgeons," *USA Today*, August 19, 2008, accessed March 7, 2013, http://usatoday30.usatoday.com/tech/gaming/2008-08-18-video-games-learning_N.htm.

18. *Ibid.*

19. Winda Benedetti, "Playing Wii Games can Help Make Doctors Better Surgeons," *NBCNews.com*, accessed March 7, 2013, http://www.nbcnews.com/technology/ingame/playing-wii-games-can-help-make-doctors-better-surgeons-1C8593122.

20. Winda Benedetti, "Want to be a Surgeon? Start Playing Video Games," *NBCNews.com*, accessed March 7, 2013, http://www.nbcnews.com/technology/ingame/want-be-surgeon-start-playing-video-games-1C7179650.

21. *Ibid.*

22. At the time, Cathleen Galas taught at the Corrine A. Seeds University Elementary School, the laboratory school for the Graduate School of Education and Information Studies at the University of California, Los Angeles (hence her connection to Dr. Kafai).

23. Cathleen Galas, "Why Whyville?," *Learning & Leading with Technology*, March 2006, 30–33.

24. Prior to Whyflu, the *Whyville* CDC office had no affiliation with the actual Centers for Disease Control and Prevention; it was simply a place for Whyvillians to go for help and information.

25. Cathleen Galas, "Why Whyville?," *Learning & Leading with Technology*, March 2006, 30–33.

26. *Ibid.*

27. Game the News, accessed March 7, 2013, http://gamethenews.net/.

28. Though these are casual games with short development and production times, a significant amount of research goes into them, and their mechanics are often complex. In *Climate Defense*, for example, the developers consulted several well-regarded sources for the science behind carbon emissions, the change in global temperature, and the resultant impact on climate (including

the National Oceanic and Atmospheric Administration, NASA, the Environmental Protection Agency, and the International Energy Agency). They then incorporated this data into the game as best they could, also factoring in the natural world's ability to absorb carbon dioxide, cost of planting trees, cost of programs to cut emissions, future cost savings of eliminating carbon from the atmosphere . . . you get the idea. For *Endgame: Syria*, the sources are even more varied (including combatants, civilians, and journalists on the ground in the heart of the conflict) and the situation, at least from a political perspective, more complicated. I won't go into them all here, but interested readers can find more details on the Game the News website, http://gamethenews.net/.

29. You can play the game online at http://www.darfurisdying.com/.

30. As of March 2013, *Half the Sky*'s partners were The Fistula Foundation, GEMS, Heifer International, ONE, Room to Read, The United Nations Foundation, and World Vision.

31. The Entertainment Software Association, *2012 Essential Facts About the Computer and Video Game Industry* (The Entertainment Software Association, 2012), 2, accessed September 5, 2012, http://www.theesa.com/facts/pdfs/ESA_EF_2012.pdf.

32. Pete Yost, "Violent Crime Plunged by 12% in U.S. in 2010," *azcentral.com*, September 17, 2011, accessed March 7, 2013, http://www.azcentral.com/news/articles/2011/09/17/20110917us-violent-crime-plunged-2010.html.

33. According to Christopher J. Ferguson, Associate Professor of Psychology and Criminal Justice at Texas A&M International University, "As a video game violence researcher and someone who has done scholarship on mass homicides, let me state very emphatically: There is no good evidence that video games or other media contributes, even in a small way, to mass homicides or any other violence among youth." Christopher J. Ferguson, "Sandy Hook Shooting: Video Games Blamed, Again," *Time*, December 2012, accessed March 7, 2013, http://ideas.time.com/2012/12/20/sandy-hook-shooting-video-games-blamed-again/.

34. Christopher J. Ferguson, Adolfo Garza, Jessica Jerabeck, Raul Ramos, and Mariza Galindo, "Not Worth the Fuss After All? Cross-sectional and Prospective Data on Violent Video Game Influences on Aggression, Visuospatial Cognition and Mathematics Ability in a Sample of Youth" *Journal of Youth and Adolescence, A Multidisciplinary Research Publication*, July 2012.

35. Lawrence Kutner and Cheryl K. Olson, *Grand Theft Childhood: The Surprising Truth About Violent Video Games, And What Parents Can Do* (New York: Simon & Schuster, 2008), 8.

36. Suzy Khimm, "POW! CRACK! What We Know About Video Games and Violence," *The Washington Post*, January 17, 2013, accessed March 7,

2013, http://www.washingtonpost.com/blogs/wonkblog/wp/2013/01/17/pow-crack-what-we-know-about-video-games-and-violence/.

37. Sharon Jayson, "Don't Study the Video Game, Study the Player," *USA Today*, September 15, 2011, accessed March 7, 2013, http://usatoday30.usatoday.com/news/health/story/health/story/2011-09-14/Dont-study-the-video-game-study-the-player/50406018/1.

38. It's interesting to note that the FBI doesn't consider video games as a cause of violence either. According to former FBI profiler Mary Ellen O'Toole, "It's my experience that video games do not cause violence. However, it is one of the risk variables when we do a threat assessment for the risk to act out violently. It's important that I point out that as a threat assessment and as a former FBI profiler, we don't see these as the cause of violence. We see them as sources of fueling ideation that's already there." David Edwards, "Former FBI Profiler: 'Video Games do not Cause Violence,'" *The Raw Story*, February 24, 2013, accessed March 7, 2013, http://www.rawstory.com/rs/2013/02/24/former-fbi-profiler-video-games-do-not-cause-violence/.

10. GAMES FOR HEALTH

1. A typical 45-minute workout for me consists of eight different stretches (four warm-up and four cool down) and 30 exercises covering cardio, strength training, upper and lower body, and core. While exercising, my heart rate averages 136 beats per minute, and I usually burn on the order of 220 calories. The exercises are fun to do, they get me working hard, and they change frequently enough to keep things interesting—thus avoiding the often mind-numbing repetition that causes people to abandon many traditional workout programs.

2. If you're unfamiliar with the game, *Dance Dance Revolution* is a music game that's played on a footpad bearing four directional arrows, one for each cardinal direction. You have to time your footwork to the beat of the music and match arrows that appear on the screen with corresponding foot stomps (front, back, left, and right—or some combination of those). You earn points for accuracy and timing. If this doesn't sound like a workout to you, then you've never seen it played at the higher (or even moderate) skill levels. The game gets *fast*. Try doing stationary foot fires for 60 seconds (about a quarter the length of the average *DDR* song), and you'll get the idea.

3. Diana L. Graf, Lauren V. Pratt, Casey N. Hester, and Kevin R. Short, "Playing Active Video Games Increases Energy Expenditure in Children," *Pediatrics*, November 25, 2008, accessed March 10, 2013, http://pediatrics.aappublications.org/content/124/2/534.full.

4. Kristina Fiore, "Exergaming Provides Real Exercise for Kids," *Medpage Today*, March 7, 2011, accessed March 10, 2013, http://www.medpagetoday.com/Pediatrics/GeneralPediatrics/25226.

5. *Ibid.*

6. Jennifer Stein, "Children Burning Calories with Video Games," *Los Angeles Times*, March 13, 2011, accessed March 10, 2013, http://articles.latimes.com/2011/mar/13/health/la-he-0313-exergaming-20110313.

7. Centers for Disease Control, "Overweight and Obesity," Accessed March 10, 2013, http://www.cdc.gov/obesity/data/adult.html.

8. *Ibid.*

9. *Ibid.*

10. Melissa Healy, "Obesity in U.S. Projected to Grow, Though Pace Slows: CDC Study," *Los Angeles Times*, May 7, 2012, accessed March 10, 2013, http://articles.latimes.com/2012/may/07/news/la-heb-obesity-projection-20120507.

11. Pam Belluck, "Children's Life Expectancy Being Cut Short by Obesity," *New York Times*, March 17, 2005, accessed March 10, 2013, http://www.nytimes.com/2005/03/17/health/17obese.html?_r=0.

12. Barbara Chamberlin, "Exergames: Time for a New Personal Best," *2012 Games For Health Conference*, Games For Health Project (2012), 9.

13. Between the time I interviewed Adam and the publication of this book, he'd left Long Island University for Yale University, leaving the ADAM Center in the quite capable hands of Dr. Shaw Bronner.

14. CBC News, "Wii Use may Aid Stroke Recovery: Study," *CBC News*, February 25, 2010, accessed March 11, 2013, http://www.cbc.ca/news/health/story/2010/02/25/stroke-rehab-wii-video-game.html.

15. *Ibid.*

16. Tolga Atilla Ceranoglu, "Star Wars in Psychotherapy: Video Games in the Office," *Academic Psychiatry*, 34, no. 3 (2010): 235, http://ap.psychiatryonline.org.

17. *Ibid.*

18. Tolga Atilla Ceranoglu, "Video Games in Psychotherapy," *Review of General Psychology*, 14, no. 2 (2010): 141–146.

19. Rhea Monique (aka Ashelia), "This Isn't the Article I Wanted to Write About *Tomb Raider*," *Hellmode*, March 21, 2013, accessed May 9, 2013, http://hellmode.com/2013/03/21/this-isnt-the-article-i-wanted-to-write-about-tomb-raider/.

20. *Ibid.*

11. WAR GAMES

1. As Colonel Scott Lambert (ret.) put it to me, "How many times can you crash a helicopter for real that costs millions of dollars? Well, you can do that once. How many times can you do it in a simulated environment?"

2. The Economist Special Report, "Video Games 'The Play's the Thing,'" *The Economist*, December 2011, 10.

3. *Ibid.*

4. Andrew Martin and Thomas Lin, "Keyboards First. Then Grenades," *New York Times*, May 2, 2011, B8.

5. Cheryl Petterin, "Army Warfighters Go Digital to Hone Skills," *US Department Of Defense*, May 10, 2011, accessed March 14, 2013, http://www.defense.gov/news/newsarticle.aspx?id=63884.

6. *Ibid.*

7. Pamela Brown, "Dismounted Soldier: Virtual Combat Training System Created for U.S. Armed Forces," *WJLA.com*, September 26, 2012, accessed March 14, 2013, http://www.wjla.com/articles/2012/09/dismounted-soldier-virtual-combat-training-system-created-for-u-s-armed-forces-80327.html. As a side note, the Dismounted Soldier Training System is also every hardcore gamer's fantasy.

8. April Dudash, "Virtual Training, Real Results," *Elite Magazine*, August 1, 2011, accessed March 14, 2013, http://fbelitemag.com/articles/2011/08/01/1105161.

9. The Economist Special Report, "Video Games 'The Play's the Thing,'" *The Economist*, December 2011, 10. Not that this is any great intellectual leap. Call me crazy, but it seems like common sense to use all available tools when tracking down such an important target in what's almost certain to be hostile territory.

10. Sue Halpern, "Virtual Iraq," *The New Yorker*, May 19, 2008, 1–6, accessed March 14, 2013, http://www.newyorker.com/reporting/2008/05/19/080519fa_fact_halpern?currentPage=1.

11. Amanda Gardner, "Depression, PTSD Plague Many Iraq Vets," *CNN.com*, June 7, 2010, accessed March 14, 2013, http://www.cnn.com/2010/HEALTH/06/07/iraq.vets.ptsd/index.html .

12. Amanda Schaffer, "Not a Game: Simulation to Lessen War Trauma," *New York Times*, August 28, 2007, accessed March 14, 2013, http://www.nytimes.com/2007/08/28/health/28game.html?_r=0.

13. The Bryant Park Project, "Recovery Aid: A Virtual Iraq, Minus the Shooting," *NPR.org*, May 20, 2008, accessed March 14, 2013, http://www.npr.org/templates/story/story.php?storyId=90858431.

14. Robert L. Hanafin, "Virtual Iraq/Afghanistan and How It is Helping Some Troops and Vets with PTSD," *Veterans Today*, July 29, 2010, accessed March 14, 2013, http://www.veteranstoday.com/2010/07/29/virtual-iraqafghanistan-and-how-it-is-helping-some-troops-and-vets-with-ptsd/.

15. Sue Halpern, "Virtual Iraq," *The New Yorker*, May 19, 2008, 1–6, accessed March 14, 2013, http://www.newyorker.com/reporting/2008/05/19/080519fa_fact_halpern?currentPage=1.

16. You may remember from Chapter 9 that the Centers for Disease Control and Prevention is experimenting with resiliency training in the hopes that it can prep its emergency personnel to deal with chaos and respond effectively in a crisis situation.

17. Jeremy Hsu, "For the U.S. Military, Video Games Get Serious," *Live Science*, August 19, 2010, accessed March 14, 2013, http://www.livescience.com/10022-military-video-games.html.

18. *Ibid.*

19. Michael Hastings, "The Drone Wars," *Rolling Stone*, April 2012, 44–47, 82.

20. *Ibid.*

21. The Air Force does not allow pilots to reveal their last names, citing what they call credible threats to their safety.

22. Elisabeth Bumiller, "A Day Job Waiting for a Kill Shot a World Away," *New York Times*, July 29, 2012, accessed March 14, 2013, http://www.nytimes.com/2012/07/30/us/drone-pilots-waiting-for-a-kill-shot-7000-miles-away.html.

23. *Ibid.*

24. *Ibid.*

25. Spiegel Online, "Interview with a Drone Pilot: 'It Is Not a Video Game,'" *Spiegel Online*, March 22, 2010, accessed March 14, 2013, http://www.spiegel.de/international/world/interview-with-a-drone-pilot-it-is-not-a-video-game-a-682842.html.

26. Michael Hastings, "The Drone Wars," *Rolling Stone*, April 2012, 44–47, 82.

12. IT'S WILLIAM GIBSON'S WORLD, WE'RE JUST LIVING IN IT

1. William Gibson, "Burning Chrome" (New York: Ace, 1987), 168–191.

2. At the time I wrote this, Google Glasses were set to be commercially available in late 2013 (at $1,500 a pair)—which means that by the time this book gets into your hands, Google Glasses will have hit the street.

3. Elizabeth Montalbano, "DARPA Works On Virtual Reality Contact Lenses," *InformationWeek*, February 1, 2012, accessed March 15, 2013, http://www.informationweek.com/government/mobile/darpa-works-on-virtual-reality-contact-l/232600054.

4. Jason Tanz, "A Thousand Points of Infrared Light," *Wired*, July 2011, 114–123.

5. Duncan Graham-Rowe, "Let the Mind Games Begin," *New Scientist*, March 2008, 40–43.

6. Stephanie Z. Pavlou "DARPA-Funded Prosthetic Arm Ready for Brain-Controlled Testing," *O&P Business News*, September 2010, accessed March 15, 2013, http://www.healio.com/orthotics-prosthetics/prosthetics/news/print/o-and-p-business-news/%7Bd1b0d410-3a75-4265-b93f-7aceafcec22c%7D/darpa-funded-prosthetic-arm-ready-for-brain-controlled-testing.

7. Science Daily, "Tactile Technology for Video Games Guaranteed to Send Shivers Down Your Spine," *Science Daily*, August 10, 2011, accessed March 15, 2013, http://www.sciencedaily.com/releases/2011/08/110808152421.htm. Of course, this sensory enhancement has other implications as well. It doesn't take a genius to see how this could apply to adult-themed games—which brings a critical issue to light: when is a game no longer a game? Right now, virtual sex is limited to (for the most part) a visual and auditory experience, and the line between the game world and reality is pretty clear—though there are issues of betrayal and trust even with that (see Chapter 8). But what happens when you incorporate touch into the game, when you can actually "feel" your game partner? And what happens if you combine this with a virtual world like *Second Life*, where your game partner could easily be a real person at the other end of a network connection? If you're single, it may not be that big a deal, but what about people who have partners in real life? Though there's no direct contact between game partners, would such immersive virtual sex be considered cheating? Or, more simply, would the betrayal be any less real?

8. Industrial Light and Magic already has highly accurate physics engines for doing complex movie rendering, but they're very slow—which, for a movie, doesn't matter, as they're not rendering movie content in real-time.

9. Rob Lindeman, personal communication.

BIBLIOGRAPHY

Activision. "*Call Of Duty®: Black Ops II* Delivers More Than $500 Million in Worldwide Retail Sales in First 24 Hours," accessed January 15, 2013, http://investor.activision.com/releasedetail.cfm?ReleaseID=721903.

Activision. "*Call of Duty®: Black Ops II* Grosses $1 Billion in 15 Days," accessed January 15, 2013, http://investor.activision.com/releasedetail.cfm?ReleaseID=725026.

Alsop II, Stewart. "TSR Hobbies Mixes Fact and Fantasy." *Inc.*, February 1, 1982, accessed February 8, 2013, http://www.inc.com/magazine/19820201/3601.html.

Ananthaswamy, Anil. "A Life Less Ordinary Offers Far More than Just Escapism," *New Scientist*, August 25, 2007.

Anderson, Mark. "The End of Gold Farming?" *IEEE Spectrum*, October 2010, accessed February 16, 2013, http://spectrum.ieee.org/consumer-electronics/gaming/the-end-of-gold-farming.

Belluck, Pam. "Children's Life Expectancy Being Cut Short by Obesity." *The New York Times*, March 17, 2005, accessed March 10, 2013, http://www.nytimes.com/2005/03/17/health/17obese.html?_r=0.

Benedetti, Winda. "Playing Wii Games can Help Make Doctors Better Surgeons," *NBCNews.com*, accessed March 7, 2013, http://www.nbcnews.com/technology/ingame/playing-wii-games-can-help-make-doctors-better-surgeons-1C8593122.

Benedetti, Winda. "Want to be a Surgeon? Start Playing Video Games," *NBCNews.com*, accessed March 7, 2013, http://www.nbcnews.com/technology/ingame/want-be-surgeon-start-playing-video-games-1C7179650.

Bhasin, Kim. "How Sundance DiGiovanni Built The Largest Competitive Gaming League in America," *Business Insider*, May 1, 2012, accessed February 3, 2013, http://articles.businessinsider.com/2012-05-01/strategy/31515986_1_esports-video-game-major-league-gaming.

Boellstorff, Tom. *Coming of Age in* Second Life. Princeton: Princeton University Press, 2008.

Bouzane, Bradley. "First-ever Virtual Surgery Performed," *Winnipeg Free Press*, August 21, 2009.

Box Office Mojo. "All Time Box Office," accessed January 15, 2013, http://www.boxofficemojo.com/alltime/.

Box Office Mojo. "Resident Evil," accessed January 15, 2013, http://www.boxofficemojo.com/search/?q=Resident%20Evil.

Brookey, Robert Alan. *Hollywood Gamers: Digital Convergence in the Film and Video Game Industries*. Bloomington: Indiana University Press, 2010.

Brown, Pamela. "Dismounted Soldier: Virtual Combat Training System Created for U.S. Armed Forces," *WJLA.com*, September 26, 2012, accessed March 14, 2013, http://www.wjla.com/articles/2012/09/dismounted-soldier-virtual-combat-training-system-created-for-u-s-armed-forces-80327.html.

The Bryant Park Project. "Recovery Aid: A Virtual Iraq, Minus the Shooting," *NPR.org*, May 20, 2008, accessed March 14, 2013, http://www.npr.org/templates/story/story.php?storyId=90858431.

Bumiller, Elisabeth. "A Day Job Waiting for a Kill Shot a World Away," *The New York Times*, July 29, 2012, accessed March 14, 2013, http://www.nytimes.com/2012/07/30/us/drone-pilots-waiting-for-a-kill-shot-7000-miles-away.html.

Caplan, Jeremy, and Ta-Nehisi Paul Coates. "Tiger. Jordan. Hawk. Wendel?" *TIME*, February 2007.

Casual Games Association. "Mobile Gaming 2012 Sector Report," accessed December 21, 2012, http://dl.dropbox.com/u/3698805/research/2012_CGA_MobileSector.pdf.

Casual Games Association. "Social Network Games 2012, Casual Games Sector Report," accessed December 21, 2012, http://dl.dropbox.com/u/3698805/research/2012_CGA_SocialSector.pdf.

Centers for Disease Control. "Overweight and Obesity," accessed March 10, 2013, www.cdc.gov.

Ceranoglu, Tolga Atilla. "Star Wars in Psychotherapy: Video Games in the Office," *Academic Psychiatry* 34 (2010): 235. Accessed March 20, 2013, http://ap.psychiatryonline.org.

Ceranoglu, Tolga Atilla. "Video Games in Psychotherapy." *Review of General Psychology* 14 (2010): 141–146.

Chamberlin, Barbara. "Exergames: Time for a New Personal Best," *2012 Games For Health Conference*, Games For Health Project, 2012.

Dibbell, Julian. "A Rape in Cyberspace," accessed March 20, 2013, http://www.juliandibbell.com/articles/a-rape-in-cyberspace/.

Dibbell, Julian. "Griefer Madness." *Wired*, February 2008.

Dibbell, Julian. "The Kingpin of Azeroth." *Wired*, December 2008.

Dibbell, Julian. *Play Money, Or, How I Quit My Day Job and Made Millions Trading Virtual Loot*. New York: Basic Books, 2006.

Dudash, April. "Virtual Training, Real Results," *Elite Magazine*, August 1, 2011, accessed March 14, 2013, http://fbelitemag.com/articles/2011/08/01/1105161.

Eberson, Sharon. "PSO Presents 'Video Games Live' to Turn on New Generation," *Pittsburgh Post-Gazette*, July 7, 2009, accessed December 6, 2012, http://www.post-gazette.com/stories/sectionfront/life/pso-presents-video-games-live-to-turn-on-new-generation-348458/.

Edwards, David. "Former FBI Profiler: 'Video Games do not Cause Violence,'" *The Raw Story*, February 24, 2013, accessed March 7, 2013, http://www.rawstory.com/rs/2013/02/24/former-fbi-profiler-video-games-do-not-cause-violence/.

Electronic Arts. "EA Announces Ultima Online (TM): Kingdom Reborn (Working Title); The Game that Firmly Established the MMORPG Genre Receives a Massive Visual Overhaul and New Content in 2007," accessed February 16, 2013, http://investor.ea.com/releasedetail.cfm?ReleaseID=314331.

eMarketer. "Mobile, Social Boost Online Gaming Populations," accessed December 21, 2012, http://www.emarketer.com/Article.aspx?R=1009100.

The Entertainment Software Association. "2011 Essential Facts About the Computer and Video Game Industry," accessed September 5, 2012, http://www.theesa.com/facts/pdfs/ESA_EF_2011.

The Entertainment Software Association. "2012 Essential Facts About the Computer and Video Game Industry," accessed September 5, 2012, http://www.theesa.com/facts/pdfs/ESA_EF_2012.pdf.

Ethic. "Interview: Istvaan Shogaatsu," *Kill Ten Rats*, October 12, 2005, accessed February 16, 2013, http://www.killtenrats.com/2005/10/12/interview-istvaan-shogaatsu/.

Fefferman Lab. "Nina H. Fefferman, PhD," accessed March 7, 2013, http://rci.rutgers.edu/~feffermn/fefferman.php.

Ferguson, Christopher J. "Sandy Hook Shooting: Video Games Blamed, Again,." *Time*, December 2012, accessed March 7, 2013, http://ideas.time.com/2012/12/20/sandy-hook-shooting-video-games-blamed-again/.

Ferguson, Christopher J., Adolfo Garza, Jessica Jerabeck, Raul Ramos, and Mariza Galindo. "Not Worth the Fuss After All? Cross-sectional and Prospective Data on Violent Video Game Influences on Aggression, Visuospatial Cognition and Mathematics Ability in a Sample of Youth," *Journal of Youth and Adolescence: A Multidisciplinary Research Publication*, July 2012.

Fiore, Kristina. "Exergaming Provides Real Exercise for Kids," *Medpage Today*, March 7, 2011, accessed March 10, 2013, http://www.medpagetoday.com/Pediatrics/GeneralPediatrics/25226.

Fox, Jesse, Jeremy N. Bailenson, and Liz Tricase. "The Embodiment of Sexualized Virtual Selves: The Proteus Effect and Experiences of Self-Objectification via Avatars." *Computers in Human Behavior* (2013): 930–938.

"Fragging on the Verge." *Game Informer Magazine*, January 2007.

Francis, Tom. "Murder Incorporated: The Story of *EVE Online*'s Most Devastating Assassins," *PC Gamer Magazine*, January 2006, archived at *ComputerAndVideoGames.com*, January 29, 2008, accessed February 16, 2013, http://www.computerandvideogames.com/100007/features/murder-incorporated/

Galas, Cathleen. "Why Whyville?" *Learning & Leading with Technology*, March 2006.

Game the News. Accessed March 7, 2013, http://gamethenews.net/.

"Game Theft Led to Fatal Attack," *BBC News*, March 31, 2005, accessed February 19, 2013, http://news.bbc.co.uk/2/hi/technology/4397159.stm

Gardner, Amanda. "Depression, PTSD Plague Many Iraq Vets," *CNN.com*, June 7, 2010, accessed March 14, 2013, http://www.cnn.com/2010/HEALTH/06/07/iraq.vets.ptsd/index.html.

Gibson, William. *Burning Chrome*. New York: Ace, 1987.

Graetz, J. M. "The Origin of Spacewar," *wheels.org*, accessed September 24, 2012, http://www.wheels.org/spacewar/creative/SpacewarOrigin.html.

Graf, Diana L., Lauren V. Pratt, Casey N. Hester, and Kevin R. Short. "Playing Active Video Games Increases Energy Expenditure in Children," *Pediatrics*, November 25, 2008, accessed March 10, 2013, http://pediatrics.aappublications.org/content/124/2/534.full.

Graham-Rowe, Duncan. "Let the Mind Games Begin," *New Scientist*, March 2008.

Halpern, Sue. "Virtual Iraq," *The New Yorker*, May 19, 2008, accessed March 14, 2013, http://www.newyorker.com/reporting/2008/05/19/080519fa_fact_halpern?currentPage=1.

Hanafin, Robert L. "Virtual Iraq/Afghanistan and How It is Helping Some Troops and Vets with PTSD," *Veterans Today*, July 29, 2010, accessed March 14, 2013, http://www.veteranstoday.com/2010/07/29/virtual-iraqafghanistan-and-how-it-is-helping-some-troops-and-vets-with-ptsd/.

Hancu, Andrei. "Angel Munoz: 'The Apex was 2004,'" *HTLV.org*, October 8, 2010, accessed February 3, 2013, http://www.hltv.org/news/5443-angel-munoz-the-apex-was-2004&nc=1.

Hastings, Michael. "The Drone Wars," *Rolling Stone*, April 2012.

Hattenstone, Simon. "The Method? Living it Out? Cobblers!" *The Guardian*, August 2, 2007, accessed January 10, 2013, http://www.guardian.co.uk/film/2007/aug/03/2.

Healy, Melissa. "Obesity in U.S. Projected to Grow, Though Pace Slows: CDC Study," *Los Angeles Times*, May 7, 2012, accessed March 10, 2013, http://articles.latimes.com/2012/may/07/news/la-heb-obesity-projection-20120507.

Herz, J. C. *Joystick Nation: How Videogames Ate Our Quarters, Won Our Hearts, and Rewired Our Minds*. Boston: Little, Brown, and Co., 1997.

Hsu, Jeremy. "For the U.S. Military, Video Games Get Serious," *Live Science*, August 19, 2010, accessed March 14, 2013, http://www.livescience.com/10022-military-video-games.html.

Huizinga, Johan. *Homo Ludens: A Study of the Play Element in Culture*. Boston: The Beacon Press, 1955.

Information Solutions Group. "2011 PopCap Games Social Gaming Research," accessed December 21, 2012, http://www.infosolutionsgroup.com/pdfs/2011_PopCap_Social_Gaming_Research_Results.pdf

International Game Developers Association. "2008-2009 Casual Games Whitepaper," accessed December 21, 2012, http://www.igda.org/sites/default/files/IGDA_Casual_Games_White_Paper_2008.pdf.

"Interview with a Drone Pilot: 'It is Not a Video Game,'" *Spiegel Online*, March 22, 2010, accessed March 14, 2013, http://www.spiegel.de/international/world/interview-with-a-drone-pilot-it-is-not-a-video-game-a-682842.html.

Jayson, Sharon. "Don't Study the Video Game, Study the Player," *USA Today*, September 15, 2011, accessed March 7, 2013, http://usatoday30.usatoday.com/news/health/story/health/story/2011-09-14/Dont-study-the-video-game-study-the-player/50406018/1.

Kafai, Yasmin B., David Feldon, Deborah Fields, Michael Giang, and Maria Quintero. "Life in the Times of Whypox: A Virtual Epidemic as a Community Event." In *Communities and Technologies*, edited by C. Steinfield, B. Pentland, M. Ackerman, and N. Contractor, 171–190. New York: Springer, 2007.

Katz, Josh. "Virtual *Maple Story* Murder Reveals Online Lives Gone Too Far," *Finding Dulcinea*, October 24, 2008, accessed February 19, 2013, http://www.findingdulcinea.com/news/technology/September-October-08/Virtual--Maple-Story--Murder-Reveals-Online-Lives-Gone-Too-Far.html.

Kent, Steven L. *The Ultimate History of Video Games: The Story Behind the Craze that Touched our Lives and Changed the World*. New York: Three Rivers Press, 2001.

Khimm, Suzy. "POW! CRACK! What We Know About Video Games and Violence," *The Washington Post*, January 17, 2013, accessed March 7, 2013, http://www.washingtonpost.com/blogs/wonkblog/wp/2013/01/17/pow-crack-what-we-know-about-video-games-and-violence/.

King, Brad, and John Borland. *Dungeons & Dreamers: The Rise of Computer Game Culture From Geek to Chic*. Blacklick: McGraw-Hill Osborne Media, 2003.

The King of Kong: A Fistful of Quarters. Directed by Seth Gordon. New Line Home Video, 2008. DVD.

Kutner, Lawrence, and Cheryl K. Olson. *Grand Theft Childhood: The Surprising Truth About Violent Video Games, And What Parents Can Do*. New York: Simon & Schuster, 2008.

Leblanc, Steve. "Studies: Video Games can Make Better Students, Surgeons," *USA Today*, August 19, 2008, accessed March 7, 2013, http://usatoday30.usatoday.com/tech/gaming/2008-08-18-video-games-learning_N.htm.

Lehrman, Robert A. "I Play Therefore I Am." *Christian Science Monitor*, March 2012.

Marriott, Michael. "Hollywood Would Kill for Those Numbers," *The New York Times*, November 14, 2004, accessed January 15, 2013, http://www.nytimes.com/2004/11/14/weekinreview/14bigp.html?_r=0.

Martin, Andrew, and Thomas Lin. "Keyboards First. Then Grenades," *The New York Times*, May 2, 2011.

Marvel. "*Avengers* Shatters More Records," accessed January 15, 2013, http://marvel.com/news/story/18690/avengers_shatters_more_records.

Mezoff, Lori. "Army Game's Medic Training Helps Save Two Lives," *Armyreal.com*, January 22, 2008, accessed March 7, 2013, http://www.armyreal.com/articles/item/3497.

Microsoft. "*Halo: Combat Evolved* for Xbox Tops 1 Million Mark in Record Time," accessed January 15, 2013, http://www.microsoft.com/en-us/news/press/2002/apr02/04-08halomillionpr.aspx.

Monique, Rhea (aka Ashelia). "This Isn't the Article I Wanted to Write About *Tomb Raider*," *Hellmode*. March 21, 2013, accessed May 8, 2013, http://hellmode.com/2013/03/21/this-isnt-the-article-i-wanted-to-write-about-tomb-raider/.

Montalbano, Elizabeth. "DARPA Works on Virtual Reality Contact Lenses," *InformationWeek*, February 1, 2012, accessed March 15, 2013, http://www.informationweek.com/government/mobile/darpa-works-on-virtual-reality-contact-l/232600054.

Morris, Steven. "*Second Life* Affair Leads to Real Life Divorce," *The Guardian*, November 13, 2008, accessed February 19, 2013, http://www.guardian.co.uk/technology/2008/nov/13/second-life-divorce

Motion Picture Association of America. "2011 Theatrical Market Statistics," accessed September 5, 2012, http://www.mpaa.org/resources/5bec4ac9-a95e-443b-987b-bff6fb5455a9.pdf.

"New Video Game Fights Teenage Depression," *New York Daily News*, August 6, 2012, accessed September 5, 2012, http://articles.nydailynews.com/2012-08-06/news/33069746_1_teenage-depression-depression-in-young-people-adolescent-depression.

Newzoo. "The Global MMO Market; Sizing and Seizing Opportunities," accessed February 16, 2013, http://www.newzoo.com/infographics/the-global-mmo-market-sizing-and-seizing-opportunities/.

Newzoo. "Online Casual & Social Games Trend Report," accessed December 21, 2012, http://www.newzoo.com/trend-reports/casual-social-games-trend-report/.

Newzoo. "Mobile Games Trend Report," accessed December 21, 2012, http://www.newzoo.com/trend-reports/mobile-games-trend-report/.

Nieva, Richard. "Video Gaming on the Pro Tour, for Glory but Little Gold," *The New York Times*, November 28, 2012, accessed February 3, 2013, http://www.nytimes.com/2012/11/[illegible]-on-the-pro-tour-for-glory-but-little-gold.html?_r=0.

Pavlou, Stephanie Z. "DARPA-Funded Prosthetic Arm Ready for Brain-Controlled Testing," *O&P Business News*, September 2010, accessed March 15, 2013, http://www.healio.com/orthotics-prosthetics/prosthetics/news/print/o-and-p-business-news/%7Bd1b0d410-3a75-4265-b93f-7aceafcec22c%7D/darpa-funded-prosthetic-arm-ready-for-brain-controlled-testing.

Petterin, Cheryl. "Army Warfighters Go Digital to Hone Skills," *U.S. Department of Defense*, May 10, 2011, accessed March 14, 2013, http://www.defense.gov/news/newsarticle.aspx?id=63884.

Quinn, Diane M., Rachel W. Kallen, and Christie Cathey. "Body on My Mind: The Lingering Effect of State Self-Objectification," *Sex Roles*, December 2006.

Robinson, Jon. "Major League Gaming Continues to Grow," *ESPN.com*, August 24, 2012, accessed February 3, 2013, http://espn.go.com/blog/playbook/tech/post/_/id/1900/major-league-gaming-continues-to-grow.

Rugg, Peter. "For Disgraced Former Joust King, There's Life After the Arcade," *The Pitch*, January 5, 2009, accessed January 23, 2013, http://www.pitch.com/kansascity/for-disgraced-former-joust-king-steve-sanders-theres-life-after-the-arcade/Content?oid=2193027&showFullText=true

Satter, Raphael G. "Virtual Affair Leads Real Divorce for UK Couple," *Huffington Post*, November 14, 2008, accessed February 19, 2013, http://www.huffingtonpost.com/2008/11/14/virtual-affair-yields-rea_n_143860.html.

Schaffer, Amanda. "Not a Game: Simulation to Lessen War Trauma." *The New York Times*, August 28, 2007, accessed March 14, 2013, http://www.nytimes.com/2007/08/28/health/28game.html?_r=0.

Schiesel, Seth. "A Game That Takes Aim at Bigger Screens," *The New York Times*, November 7, 2009.

Schiesel, Seth. "Brave New World That's as Familiar as the Machine It Fights With," *The New York Times*, July 6, 2011.

Seitz, Holli. "Whyville and the 2009-2010 Whyflu: Evolving a Virtual World Activity to Meet Changing 'Real World' Communication Needs." (Abstract). *Convergence 2010: National Conference on Health, Communication, Marketing, and Media*, accessed March 7, 2013, https://cdc.confex.com/cdc/nphic10/webprogram/Paper24781.html.

Siwek, Stephen P. "Video Games in the 21st Century: The 2010 Report," *The Entertainment Software Association*, 2010, accessed September 5, 2012, http://www.theesa.com/facts/pdfs/VideoGames21stCentury_2010.pdf.

Simple English Wikipedia. "Leet," last modified January 2013, http://simple.wikipedia.org/wiki/Leet.

Sorkin, Andrew Ross. "Angry Birds Maker Posted Revenue of $106.3 Million in 2011," *New York Times Dealbook*, May 7, 2012, accessed December 21, 2012, http://dealbook.nytimes.com/2012/05/07/angry-birds-maker-posts-2011-revenue-of-106-3-million/.

Stark, Chelsea. "Move Over, Super Bowl. Spectator Gaming Reaches Millions Online," *Mashable.com*, February 27, 2012, accessed February 3, 2013, http://mashable.com/2012/02/27/spectator-gaming-esports/.

Stein, Jennifer. "Children Burning Calories with Video Games," *Los Angeles Times*, March 13, 2011, accessed March 10, 2013, http://articles.latimes.com/2011/mar/13/health/la-he-0313-exergaming-20110313.

Steinkuehler, Constance, and Sean Duncan. "Scientific Habits of Mind in Virtual Worlds," *Journal of Science Education and Technology* 17 (2008): 530–543.

"Tactile Technology for Video Games Guaranteed to Send Shivers Down Your Spine," *Science Daily*, August 10, 2011, accessed March 15, 2013, http://www.sciencedaily.com/releases/2011/08/110808152421.htm.

Tanz, Jason. "A Thousand Points of Infrared Light," *Wired*, July 2011.

Tassi, Paul. "Talking the Past, Present and Future of eSports with MLG's Sundance DiGiovanni," *Forbes*, August 22, 2012, accessed February 3, 2013, http://www.forbes.com/sites/insertcoin/2012/08/22/talking-the-past-present-and-future-of-esports-with-mlgs-sundance-digiovanni/.

Taub, Eric A. "Taking Their Game to the Next Level," *The New York Times*, October 7, 2004, accessed February 3, 2013, http://www.nytimes.com/2004/10/07/technology/circuits/07play.html?pagewanted=1&_r=0.

Thier, David. "Summer 2010: When Video Games Took Over the Movies," *The Atlantic*, August 2010, accessed January 10, 2013, http://www.theatlantic.com/entertainment/archive/2010/08/summer-2010-when-video-games-took-over-the-movies/61522/.

Thier, David. "*World of Warcraft* Shines Light on Terror Tactics," *Wired*, March 2008, accessed March 7, 2013, http://www.wired.com/gaming/virtualworlds/news/2008/03/wow_terror.

Thompson, Anne. "What Will *Tron: Legacy*'s 3D VFX Look Like in 30 Years?" *Popular Mechanics*, December 2010, accessed January 10, 2013, http://www.popularmechanics.com/technology/digital/visual-effects/are-tron-legacys-3d-fx-ahead-of-their-time.

Thompson, Clive. "The Science of Play." *Wired*, September 2007.

Townsend, Emru. "The 10 Worst Games of All Time," *PC World*, October 2006, accessed January 15, 2013, http://www.pcworld.com/article/127579/article.html?page=2.

Video Games Live Level 2. Directed by Allen Newman. Sherman Oaks, CA: Mystical Stone Entertainment, 2010. DVD.

"Video Games: 'The Play's the Thing,'" *The Economist*, December 2011.

"Virtual Theft Leads to Arrest," *BBC News*, November 14, 2007, accessed February 19, 2013, http://news.bbc.co.uk/2/hi/technology/7094764.stm.

Warman, Peter. "The Bigger Picture: Key Facts and Trends on the Ever-changing Global Games Market." Keynote presentation delivered at LU-CIX conference, Luxembourg, September 2012.

"Wii Use may Aid Stroke Recovery: Study," *CBC News*, February 25, 2010, accessed March 11, 2013, http://www.cbc.ca/news/health/story/2010/02/25/stroke-rehab-wii-video-game.html.

Wikipedia. "*E.T. the Extra-Terrestrial* (Video Game)," last modified January 2013, http://en.wikipedia.org/wiki/E.T._the_Extra-Terrestrial_(video_game).

Wikipedia. "*Halo: Combat Evolved*," last modified January 2013, http://en.wikipedia.org/wiki/Halo:_Combat_Evolved#Reception.

Wikipedia. "List of *Halo* Media," last modified January 2013, http://en.wikipedia.org/wiki/List_of_Halo_media.

Wilhelm, Alex. "How Major League Gaming went All-in, Landed on Its Feet, and Raised Millions," *TNW: The Next Web*, March 13, 2012, accessed February 3, 2013, http://thenextweb.com/insider/2012/03/13/how-major-league-gaming-went-all-in-landed-on-its-feet-and-raised-millions/.

Wilmington, Michael. "No Offense Nintendo: *Super Mario Bros.* Jump to Big Screen in Feeble Extravaganza," *Los Angeles Times*, May 29, 1993, accessed January 10, 2013, http:/ /articles.latimes.com/1993-05-29/entertainment/ca-41093_1_super-mario-bros.

World Cyber Games. "WCG Concept," accessed February 3, 2013, http://www.wcg.com/ renew/inside/wcgc/wcgc_concept.asp.

WR Hambrecht+Co. "The U.S. Professional Sports Market & Franchise Value Report, 2012," accessed September 5, 2012, http://www.wrhambrecht.com/pdf/SportsMarket Report_2012.

Yee, Nick, and Jeremy Bailenson. "The Proteus Effect: The Effect of Transformed Self-Representation on Behavior," *Human Communication Research* (2007): 271–290.

Yee, Nick, Jeremy N. Bailenson, and Nicolas Ducheneaut. "The Proteus Effect: Implications of Transformed Digital Self-Representation on Online and Offline Behavior," *Communication Research* (2009): 285–312.

Yost, Pete. "Violent Crime Plunged by 12% in U.S. in 2010," *azcentral.com*, September 17, 2011, accessed March 7, 2013, http://www.azcentral.com/news/articles/2011/09/17/2011 0917us-violent-crime-plunged-2010.html.

INDEX

ABOUT THE AUTHOR

Devin C. Griffiths has been writing all his life and gaming almost as long. He grew up in Tarrytown, New York, during the great video game boom of the late '70s and early '80s and spent many hours (and more quarters) in their company. He studied science journalism at Hampshire College and launched his own PR and marketing company, Catamount Communications, in the early 2000s. He lives in western Massachusetts with his wife and son and is still an avid gamer. This is his first book.

BERKELEY COLLEGE LIBRARY/WESTCHESTER
3 6615 4002 3673 6